Cultural Heritage, Community Engagement and Sustainable Tourism

This timely and innovative book critically explores how cultural heritage in the Global South can be used to mobilise community engagement and promote sustainable tourism at archaeological sites.

Whilst the volume covers theoretical issues, it primarily offers insight into how both small and large projects within low- and middle-income countries start, plan and develop. It describes what factors lead some projects to succeed, some to fail and most to have elements of both. Core to this investigation, each specifically commissioned chapter considers the challenges of developing collaboration and joint ownership between multiple stakeholders, ranging from local communities to national governments. In addition, the book considers how the challenges of the COVID-19 pandemic were managed, with lessons for future events of such magnitude. In summary, this significant volume recognises the value of developing collaborative partnerships between academics, NGOs and local communities, to achieve community engagement within archaeological research and support sustainable development by developing appropriate forms of tourism at archaeological sites. It provides essential reading for those interested in tourism, heritage studies, archaeology, geography, tourism studies and cultural studies.

Steven Mithen, FBA, is Professor of Early Prehistory at the University of Reading, UK. His research interests include the origins of language, music and thought, hunter-gatherers and the emergence of farming. He has undertaken long-term field projects in southern Jordan and western Scotland and is a founding member of a local charity in Scotland (Islay Heritage, SCO46938). His books include *The Prehistory of the Mind* (1996), *After the Ice: A Global Human History 20,000–5000 BC* (2003), *The Singing Neanderthals* (2005), *Thirst: Water and Power in the Ancient World* (2007) and *The Language Puzzle* (2024).

Mubariz Ahmed Rabbani is working as a research engineer in archaeology at the French National Centre for Scientific Research (CNRS). His research interests include the study of the archaeology of South Asia and South-West Asia, focusing on ancient technologies, socio-economic and political organisation, trade, religion and human–environmental relationships. He completed his PhD in archaeology at the University of Reading and has worked on excavations in Pakistan, Iraq and the UK. He is Member of the ISMEO Italian Archaeological Mission in Pakistan.

Maria Rabbani is a Palynologist for Oxford Archaeology. She completed her PhD at the University of Reading, which focused on human-environmental interactions in the Zagros region during the Late Pleniglacial, Lateglacial and Holocene, using pollen, non-pollen palynomorph, micro- and macro-charcoal and geochemical analyses. Maria has experience in working on lake and wetland sediment and pollen from the UK, Italy, Iran and Iraq.

Contemporary Geographies of Leisure, Tourism and Mobility
Series Editor: C. Michael Hall, *Professor at the Department of Management, College of Business and Economics, University of Canterbury, Christchurch, New Zealand*

The aim of this series is to explore and communicate the intersections and relationships between leisure, tourism and human mobility within the social sciences.

It will incorporate both traditional and new perspectives on leisure and tourism from contemporary geography, e.g. notions of identity, representation and culture, while also providing for perspectives from cognate areas such as anthropology, cultural studies, gastronomy and food studies, marketing, policy studies and political economy, regional and urban planning and sociology, within the development of an integrated field of leisure and tourism studies.

Also, increasingly, tourism and leisure are regarded as steps in a continuum of human mobility. Inclusion of mobility in the series offers the prospect to examine the relationship between tourism and migration, the sojourner, educational travel and second home and retirement travel phenomena.

The series comprises two strands:

Contemporary Geographies of Leisure, Tourism and Mobility aims to address the needs of students and academics, and the titles will be published simultaneously in hardback and paperback.

Routledge Studies in Contemporary Geographies of Leisure, Tourism and Mobility is a forum for innovative new research intended for research students and academics, and the titles will initially be available in hardback only. Titles include:

Travel Disruptions
Impacts, Responses and Resilience
Edited by Emmet McLoughlin, Domhnall Melly and James Hanrahan

The Business of Poverty in Africa
Inside the Travel Philanthropy System
Amy Scarth and Marina Novelli

Cultural Heritage, Community Engagement and Sustainable Tourism
Case Studies from Archaeological Sites in the Global South
Edited by Steven Mithen, Mubariz Ahmed Rabbani and Maria Rabbani

For more information about this series, please visit: www.routledge.com/Contemporary-Geographies-of-Leisure-Tourism-and-Mobility/book-series/SE0522

Cultural Heritage, Community Engagement and Sustainable Tourism

Case Studies from Archaeological Sites in the Global South

Edited by Steven Mithen, Mubariz Ahmed Rabbani and Maria Rabbani

Routledge
Taylor & Francis Group
LONDON AND NEW YORK

First published 2025
by Routledge
4 Park Square, Milton Park, Abingdon, Oxon OX14 4RN

and by Routledge
605 Third Avenue, New York, NY 10158

Routledge is an imprint of the Taylor & Francis Group, an informa business

British Library Cataloguing-in-Publication Data
A catalogue record for this book is available from the British Library

ISBN: 9781032792255 (hbk)
ISBN: 9781032792262 (pbk)
ISBN: 9781003491071 (ebk)

DOI: 10.4324/9781003491071

Typeset in Times New Roman
by Newgen Publishing UK

Contents

Figures

Contributors

Tohamy Abulgasim is a research assistant with the 'Mapping Africa's Endangered Sites and Monuments' project based at the University of Cambridge. He is also Co-Director of Community Engagement for the Jebel Barkal Archaeological Project with the University of Michigan. He has worked on numerous archaeological projects throughout Sudan.

Mario Advíncula is an archaeologist and cultural heritage manager. He is an archaeology graduate, specialising in cultural rights and management, with postgraduate diplomas in museology, change management and territorial development. He is Community Liaison Officer for the *Past Water Futures* Project, dedicated to re-evaluating the present-day heritage and functional value of pre-Hispanic dams in the Andes.

Yusra Al-Jabari is Project Manager and Project Monitoring and Evaluation Specialist for Future Pioneers for Empowering Communities in Environmental and Educational Fields (FPEC). She graduated from the Arab Open University. Before joining FPEC, Yusra became the Project Education Officer in Mafraq, Jordan, leading a team to oversee the delivery of educational services in vulnerable communities.

Fatima Al-Nammari obtained her PhD through a Fulbright Grant from Texas A&M University. She is Dean of the Faculty of Architecture and Design and Head of the Department of Architecture at Petra University, Chair of the Scientific Committee in the Architectural Division of the Jordan Engineers Association and Chair of the Consultancy Committee and Member of the Administrative Board of the Jordan Architects Society. She has worked in the fields of participatory urban design and planning, disaster preparedness and heritage conservation. She is Honorary Research Fellow at the University College, London and Co-winner of the Newton Prize 2020 with Professor Steven Mithen.

Fathy Awad is an Egyptologist and archaeologist. He is Director General of the South Minia Archaeological Region for the Egyptian Ministry of Tourism and Antiquities. He has more than 25 years of experience in fieldwork and site management across Egypt, from Luxor to Alexandria and has led excavations at Tuna el-Gebel and other sites. He is an active researcher, with recent publications on the archaeology of Tuna el-Gebel and contributions to both the Tell el-Amarna Site Management Plan 2020 and the book *Amarna: A Guide to the Ancient City of Akhetaten*.

Grace Barretto-Tesoro is Professor at the University of the Philippines School of Archaeology and earned her PhD from the University of Cambridge. Her research focuses on mortuary practices, identity, cosmology and historical archaeology. She is involved in the Pantropocene project, edited *Exploring Philippine Cemeteries* (2016) and authored *Identity and Reciprocity in 15th Century Philippines* (2008) and *The Ruins of Pinagbayanan* (2021). She is a member of the Society of Philippine Archaeologists.

Daniele Morandi Bonacossi is Professor of Near Eastern Archaeology at the University of Udine. He has received a PhD from the Università L'Orientale, Naples. He has been a research fellow at the Free University of Berlin and the University of Munich. Daniele has directed many campaigns of archaeological excavation and survey in various sites and regions of West Asia (Syria, Oman, Yemen, Iraq). He leads the Italian Archaeological Mission to the Kurdistan Region of Iraq (Land of Nineveh Archaeological Project).

Rebecca Bradshaw is an affiliated scholar at the University of Cambridge's MacDonald Institute for Archaeological Research. She is also Academic Affiliate with the University of Michigan, for whom she is Co-Director of Community Engagement for the Jebel Barkal Archaeological Project (2020–present). She also writes for news outlets such as *Al-Monitor* and presents historical documentaries for National Geographic.

Nicholas Branch is Professor in Palaeoecology at the University of Reading in the Department of Geography and Environmental Science. He has a PhD in Geography from Royal Holloway, University of London. His research focuses on environmental change in the Peruvian Andes. He has specialist expertise in the application of palaeoecological and geochronological methods.

Rebecca Bria is an anthropological archaeologist whose research in the Peruvian Andes examines the practices through which ancient people shaped environments and communities, focused on the intersection of food production and ritual practice. She works closely with Quechua-speaking communities on collaborative research endeavours and a variety of co-creative initiatives. Bria serves as Assistant Professor of Anthropology at the University of Texas at San Antonio. She established and co-directs the *Proyecto de Investigación Arqueológico Regional Ancash* or PIARA archaeological project.

Alanna Cant is Lecturer in Social Anthropology at the University of Reading and Author of *The Value of Aesthetics: Oaxacan Woodcarvers in Global Economies of Culture* (2019). Her research looks at themes in religion, value, aesthetics, property and materiality in the culture and heritage sectors of Mexico and the United Kingdom.

Manuel Castro-Priego is Professor in the Department of History and Philosophy, Area of Archaeology, Universidad Alcalá, Madrid, Spain. He has more than 40 publications including books and scientific articles. He has participated in multiple investigations in Europe, Africa and Latin America. He is Co-Director of the research project 'Perdurations, continuity and rupture. New realities of inequality on the central Ecuadorian coast (16th–17th centuries): archaeological indicators and environmental transformations' established in Ligüiqui, Ecuador.

Elmontaser Dafalla currently works as Senior Conservator at Sudan's National Corporation for Antiquities and Museums (NCAM). Over the past 20+ years, he has worked with numerous international missions; more recently, he has participated in the restoration of museum collections and archaeological sites. Dafalla has lately finished his doctorate at the University of Adam Mickiewicz in Poland.

Suzanne Davis specialises in the conservation of archaeological heritage. She currently heads the Conservation Department at the University of Michigan's Kelsey Museum of Archaeology and is President of the American Institute of Conservation. She works with the International Kurru Archaeological Project and the Jebel Barkal Archaeological Project. Her research and publications focus on archaeological conservation and community-engaged practice.

Sami Elamin is Director of the Northern State Antiquities Office for Sudan's National Corporation for Antiquities and Museums (NCAM), Director of the NCAM Heritage Register and Archive Unit and the Resident Manager of Jebel Barkal and the Napatan Region World Heritage Properties. He has worked on several archaeological projects in the region since 2013, most recently with the NCAM-University of Michigan team at El-Kurru and at Jebel Barkal since 2017.

Geoff Emberling is an archaeologist and museum curator specialising in North Africa and the Middle East, where his research interests include ancient urbanism, empires and identities in Mesopotamia and the Nile Valley. His interest in community engagement and collaboration, facilitated by his experience in museums, began with his work in the village of El-Kurru in 2016. Among his recent research publications are the *Oxford Handbook of Ancient Nubia* (co-edited with Bruce Williams, 2021) and an overview of the Napatan period of ancient Kush in the *Oxford History of the Ancient Near East*, v. 4 (2023).

Bill Finlayson is the Director of the Endangered Archaeology of the Middle East and North Africa (EAMENA) project. Initially trained as an archaeologist at Edinburgh University, Bill has worked in cultural heritage management since 1990, when he helped establish the Centre for Field Archaeology. He has co-established the Endangered Cultural Heritage in the Global South (ECHGS) Hub at Oxford.

Pedro Gonzalez is Conservation Architect with a B.A. in Architecture and further studies in conservation and restoration. He is a scion of a family of traditional artists from the Mantaro Valley. He is Conservator of the *Qhapaq Ñan* Project for the Vilcashuamán Sangalla section and Head Conservator of the *Past Water Futures* Project, restoring pre-Hispanic dams in the Andean region.

Josephine Handley is a postdoctoral research associate focusing on past human–environmental interactions. She has a PhD in Archaeology and a MSc in Environmental Archaeology from the University of Reading. Her PhD thesis researched the socio-economic responses to Late Holocene climate and environmental change in the Peruvian Andes.

Bekas Jamaluddin Hasan is the Duhok Director of Antiquities and Heritage (Kurdistan Region of Iraq). He got his PhD in Archaeology at the University of Lyon and he is a specialist in ancient rock reliefs from the Kurdistan Region of Iraq. He is involved in several projects concerning Heritage Protection and Management in the Duhok Governorate. Dr Bekas is Co-Director of the Kurdish-Italian Faida Archaeological Project (KIFAP) and of the Kurdish-Italian Gomel Archaeological Project (KIGAP). He is committed to raising awareness on Heritage Protection.

Alexander Herrera is Associate Professor at the Faculty of Arts and Humanities of Universidad de los Andes, Colombia. He is a Fellow of the Society of Antiquaries of London, Research Associate of the Institute of Andean Studies and a member of ICOMOS Colombia and the Information and Discussion Network on Archaeology and Heritage (RIDAP). He has conducted wide-ranging research in the central and northern Andes.

Obyda Hummash is the CEO of Future Pioneers for Empowering Communities in Environmental and Educational Fields (FPEC). She is Certified Project Management Professional and Board Member of the Project Management Institute (PMI) Jordan Chapter. At FPEC, Obyda manages a wide range of development projects.

Jessica Jacob works on heritage and tourism in the Middle East. She is interested in how heritage is visualised, remembered and enacted through the production of tourist space. Jessica is Founder of *Film Geographies*, an online forum founded in 2016 for films and filmmaking as a form of academic practice and knowledge production. Her current projects include *Storytelling for All: Skillshare Bedouin Women and Men through Film and Craft Impact and Engagement* that will enhance the value of Bedouin tangible and intangible cultural heritage in South Sinai, Egypt, through the production of a handcrafted community map wall hanging.

Elaine Jamieson is a specialist in landscape archaeology, with a particular interest in how people in the past perceived natural and cultural landscapes, and the importance of place in constructing and maintaining collective and personal identities. In 2015, Elaine joined the University of Reading where she gained her PhD in Archaeology. She joined the University of York in 2022 and currently leads the Skipsea Landscape Project.

Juan Andrés Jijón-Porras works in archaeological research activities in Ecuador, with interventions in both preventive and rescue archaeology. He is currently Research Professor at the Faculty of Humanistic and Social Sciences, Technical University of Manabí, Manabí, Ecuador. He works as Coordinator of Cultural Heritage and Social Memory of the Municipality of Portoviejo. He is General Manager of ArqueoSapiens S.A., a company dedicated to the management of cultural heritage with an emphasis on archaeology.

Hamada Kellawy is an Egyptologist and archaeologist, who holds a Master's degree in Archaeology from the University of Minia. He has participated in many fieldwork projects in Egypt, including at Amarna, in the Valley of the Kings and on the Giza plateau. He is currently Chief Inspector of Southern Malawi Antiquities for the Egyptian Ministry of Tourism and Antiquities and is involved in the day-to-day management of multiple pharaonic heritage sites. His book *Amarna: A Guide to the Ancient City of Akhetaten* was recently published.

Kevin Lane is Lecturer at the University of Buenos Aires and Senior Researcher at CONICET (National Scientific and Technical Research Council of Argentina). He received his DPhil in Archaeology at the University of Cambridge (2006). He is Director of the *Past Water Futures* project dedicated to the rehabilitation of pre-Hispanic dams for present-day use. He specialises in historical archaeology and heritage management.

Paul Lane is the Jennifer Ward Oppenheimer Professor of the Deep History and Archaeology of Africa at the University of Cambridge. He began researching Africa as a PhD student and over the last five decades or so has conducted archaeological and/or ethnoarchaeological research in Kenya, Tanzania, South Sudan, Mali, Botswana, Senegal and Mozambique. His research interests include the transitions to food production in eastern and southern Africa, the maritime archaeology of the Indian Ocean coast and the archaeology of pastoralism in East Africa.

Caroline Marie Q. Lising is a university lecturer, archaeologist, entrepreneur and the Honorary Consul of Georgia in Manila. She received her PhD from Goethe University in Frankfurt (Magna cum laude), focusing on prehistoric archaeological heritage management in Cagayan Valley, Philippines, where she is from. She has contributed to research published in *Nature* and *Radiocarbon*. She is a faculty member at Ateneo de Manila University and a guest researcher with the ROCEEH Program of the Senckenberg Research Institute in Germany. Caroline founded *Crania Heritage Sciences*, Inc. and manages *One Fort Santiago* (OFS).

Noel Lwoga is the CEO of the National Museum of Tanzania. He has 15 years of experience as a heritage and tourism academic, researcher and strategist interested in how people perceive and experience heritage and leverage its synergy with tourism business to enhance conservation action, economic growth and sustainable communities. He is a senior research associate at Johannesburg University and a member of Tanzania's National Tourism Working Group and

National Cultural and Arts Fund. He holds a PhD in Business Administration and Tourism.

Nebras Maslamani is a certified Project Management Professional and Environmental Engineer. Following a degree from the German Jordanian University in Jordan, Nebras took an MBA from South Valley University in Egypt. She has worked in Jordan, Morocco, Germany and the United States, with a focus on ethical management, hands-on expertise and cultural adaptability. Nebras is currently based in Saudi Arabia focusing on enhancing employee productivity, streamlining processes and fostering a culture of innovation.

Frank Meddens is a visiting research fellow at the School of Archaeology, Geography and Environmental Science of the University of Reading. He is a fellow of the Society of Antiquaries of London, a member of the Chartered Institute for Archaeologists and a research associate of the Institute of Andean Studies and has conducted wide-ranging research in Britain and in the Peruvian Andes.

Elinaza Mjema is Senior Lecturer in Archaeology at the University of Dar es Salaam and Head of the Department of Archaeology and Heritage Studies. His research focuses on exploring change and continuity in ancient settlements on the western Indian Ocean coast, through the analysis of ceramics, beads and faunal remains. He is currently developing a science dissemination project using mobile museum education tools to communicate information about past climate changes and their relevance to understanding the current climate crisis.

El Hassan Ahmed Mohamed has participated in several national and international archaeological missions in Sudan. He was Co-Director of the Jebel Barkal Archaeological Mission (1996–2022). Though he retired in December 2022, he has continued working at Jebel Barkal and is a member of NCAM's Crisis Response Committee in Cairo. El Hassan is the author of several books including *King Taharqa 699-664 B.C: His Life and Monuments in the Nile Valley* (in Arabic, 2020) and *The Complete Translation of the Kushite Kings' Stelae* (in Hieroglyphs, Arabic and English, 2024) and co-author of *Sudan's Holy Mountain: Jebel Barkal and its Temples*.

Wilmer Moncada is Professor and RENACYT Researcher at the Department of Mathematics and Physics of the Faculty of Mining, Geology and Civil Engineering of the National University of San Cristobal de Huamanga – UNSCH. He is a specialist in remote sensing techniques, geographic information systems and hydrological modelling.

Luca M. Olivieri (ISMEO/Ca' Foscari) is the Director of the Italian Archaeological Mission in Pakistan (Swat) and Scientific Co-Director of the Italian Archaeological Mission in Sindh (Banbhore). He has directed conservation, excavation and survey projects in northern Pakistan (from Swat to Baltistan) and reconnaissance projects in India (Kerala and Tamil Nadu) and has been a UNESCO consultant for projects in Afghanistan (Logar and Bamiyan provinces). Luca has published over 220 articles and monographs.

Carol Palmer is a British anthropologist, environmental archaeologist and botanist. She completed her PhD at Sheffield University. Carol served as the Director of the CBRL's British Institute in Amman and the overall Director of CBRL. Her primary research interests are in rural societies in the Arab world, changes in food production practices on the landscape and in society and ethnobotany. She was part of the following projects: 'Identifying activity areas in Neolithic sites through ethnographic analysis of phytoliths and geochemical residues' (INEA), 'Mapping Digital Cultural Heritage in Jordan' (MaDiH), Antikythera Survey and Wadi Faynan Landscape Survey.

Fernando Represa-Pérez is Chair of Anthropology and Full-Time Principal Professor-Researcher at the Universidad Laica Eloy Alfaro of Manabí (Ecuador). He is Principal Investigator of the Interculturality and Biocultural Heritage Group (ULEAM). His research focuses on religiosity, cultural heritage and human-environmental relationships in marine–coastal contexts.

Maria Elena Ronza is an architect, archaeologist and President of Sela, an organization for the active engagement of Jordanians in the protection of their cultural heritage. Her career focuses on sustainable tourist development of archaeological and heritage sites through community engagement. Currently, Sela is engaged with the Petra Development and Tourism Regional Authority in the maintenance of the WHS of Petra through the engagement of trained community members.

Gehan Selim is the Hoffman Wood Professor of Architecture and Deputy Director at Leeds Social Sciences Institute at the University of Leeds. She obtained a PhD in Architecture from the University of Sheffield. She serves on the editorial board and scientific committee of several international journals and conferences. Dr Selim is an affiliate member of the Royal Institute of British Architects (RIBA) and an associate member of the Higher Education Academy (HE).

Francesca Simi is a postdoctoral researcher at the University of Udine and the Deputy Director of the Land of Nineveh Archaeological Project (LoNAP) and the Sennacherib's Archaeological Park (SAP) project. She obtained her PhD in Archaeology jointly from the Universities of Venice and Tübingen. Francesca is an archaeologist of Southwest Asia specialising in Landscape Archaeology. Her interests focus on the reconstruction of long-term population and land-use dynamics in Upper Mesopotamia. She has been involved in several survey and excavation projects in Syria, Turkey, Lebanon, Iraq and Italy.

Kate Spence Senior Lecturer in Egyptian Archaeology at the University of Cambridge and Fellow of Emmanuel College. She was UK Principal Investigator of the Amarna Institutional Links project discussed in this volume. She is a specialist in settlement archaeology, architecture and the Amarna period. She has excavated in the Central City and North Palace at Amarna. Her most recent fieldwork has focused on the ancient Egyptian colonial town of Sesebi in Northern Sudan, a site almost exactly contemporary with Amarna.

Anna Stevens is Lecturer in Archaeology at Monash University, Australia and Honorary Research Associate at the McDonald Institute for Archaeological Research, University of Cambridge, UK. She has been a member of the research team to Amarna for over two decades. She is currently overseeing a multidisciplinary initiative to investigate best-practice approaches to recording and protecting mudbrick houses excavated at Amarna in the early 1900s. Her publications include *Busy Lives at Amarna: Excavations in the Main City* (with B. Kemp), *Akhenaten's Workers: The Amarna Stone Village Survey*, 2005–2009, and, as editor, *Amarna: A Guide to the Ancient City of Akhetaten*.

Gemma Tully is an independent community archaeologist and museum professional who specialises in collaborative and co-production projects in Egypt, Sudan and Europe. She was Research Associate at Cambridge University for the Institutional Links Amarna heritage project. She has published widely in the fields of archaeology and heritage management and has taught museum studies and public archaeology courses. Her most recent museum role involved co-ordinating the 'Tutankhamun the Boy: Growing up in Ancient Egypt' project at the Petrie Museum of Egyptian and Sudanese Archaeology, UCL and she has recently become an advisor to the National Trust.

Erick Casanova Vásquez is an archaeologist with more than ten years of experience conducting archaeological research across the Peruvian Andes. Erick codirects the *Proyecto de Investigación Arqueológico Regional Ancash*, or PIARA archaeological project, which promotes initiatives that encourage heritage conservation, cultural identity and community development in Peru.

Cirilo Vivanco is a *Llaqta runa* from the *Qullana* ayllu, native to the Andean town of Sarhua, Ayacucho, Peru. He is a senior researcher at the National University of San Cristóbal de Huamanga (UNSCH) and a member of the National Scientific, Technological and Technological Innovation Registry (RENACYT). He is interested in Late Intermediate and Late Horizon period archaeology.

Douglas Walsh studied zoology at Royal Holloway College, before working as Archaeologist for the Museum of London. He recently worked as Peruvian Representative for the University of Reading on projects to investigate adaptations to climate change in the past, present and future in rural areas of Peru in Ancash, Apurimac and Ayacucho.

Stephanie Wynne-Jones is Professor of African Archaeology at the University of York. She has more than 20 years of experience working on the Swahili coast of eastern Africa and has led several excavation projects across the region. Her work includes a substantial community component, and she has published widely on the heritage and community archaeology of Tanzania. Professor Wynne-Jones is also Honorary Research Fellow at the University of South Africa (UNISA).

Leonardo Alonzo-Zambrano is a tourism engineer and Master in Tourism Business Management from the Universidad Laica Eloy Alfaro de Manabí, Ecuador. He is an activist for cultural, natural and community rights and has worked as a local guide for more than 15 years. He is an official of the Ministry of the Environment, Water and Ecological Transition of Ecuador and works as Park Ranger of the Pacoche Wildlife, Marine and Coastal Refuge.

1 Introduction

Steven Mithen, Mubariz Ahmed Rabbani and Maria Rabbani

This volume describes fourteen projects at archaeological sites in the Global South that have sought to engage with and benefit the communities living in the vicinity of those sites, often by seeking to support sustainable heritage-based tourism. The sites range from small Neolithic settlements to colonial architecture, from boulder-built fishing pens to huge, monumental sculptures cut into rock faces. All have either suffered from partial destruction or face threats to their future, and usually both; they are in middle- or low-income countries, frequently within areas of arresting economic impoverishment. Moreover, all sites are of considerable academic significance, requiring not only conservation but also further research to address globally relevant issues about human adaptability and culture change. As such, this volume contributes to two overlapping areas of burgeoning literature: the role of heritage for sustainable economic development (e.g. Baillie and Sørensen, 2021; Cernea, 2001; Giliberto and Labadi, 2022; Gould and Burtenshaw, 2014; Hampton, 2005; Labadi, 2022; Labadi et al., 2020, 2021; Madandola and Boussaa, 2023) and the nature of archaeological practice in a post-colonial, ecologically fragile, economically unequal and conflict-ridden world (e.g. Chirikure, 2021; Phillips and Allen, 2010; Waterton and Watson, 2011).

In '*Rethinking Heritage for Sustainable Development*' (2022), Sophie Labadi argues that cultural heritage, whether tangible or intangible, can be utilised in pursuit of the United Nations Sustainable Development Goals (SDGs). While she provides a compelling argument, the immense complexities, challenges and pitfalls of achieving such ends are exposed by her critical assessment of four international heritage projects implemented in sub-Saharan Africa. Each project had multimillion-dollar funding matched by a wealth of expertise and initiatives. While differing in many ways, one common approach had been to support sustainable, heritage-based tourism: that which delivers economic benefit directly to local communities rather than to international operators and does so with minimal environmental cost.

Labadi evaluates the projects on their contributions to poverty reduction, gender equality and environmental sustainability. As she explains:

DOI: 10.4324/9781003491071-1

> understanding what happened in these projects, their legacy, and what they can teach us in terms of design, management and implementation is very important … learning lessons from the past through critical analyses ensures we avoid repeating mistakes in the future.
>
> (Labadi, 2020, p. 5)

Our aim as editors of this volume is not dissimilar, although our case studies are smaller in scale (with perhaps two exceptions), and we leave our authors to reflect on the strengths and weaknesses of their own projects, with an invitation to readers to join them in doing so.

Few of us have the desire, opportunity and ability to lead projects at the scale that Labadi describes. Many academics, however, are attracted to the idea of using heritage for the benefit of local communities, whether as the primary focus of their career or as one task within a portfolio of interests and activities. Similarly, few of us have the mental bandwidth to cover the entirety of 'heritage', being focussed on how one specific component of that unbounded category, whether it be music, food, architecture or – as in this book – archaeology, might be used for the public good.

In the UK, and we suspect elsewhere, archaeologists are increasingly being pushed in the direction of supporting SDGs by our funding bodies and ultimately the government behind them. They wish to see our research delivering 'impact', often defined as enhancing social well-being and/or contributing to economic growth. Unless we embrace impact, funding for our research will be limited, if available at all.

Whether or not seeking the lofty goals of poverty alleviation, gender equality or other SDG themes, there is a compelling case that archaeological projects need to engage with those communities living in the geographical proximity of the sites we survey and excavate. A great many projects do so, especially within the relatively prosperous regions of the Global North where this is facilitated by networks of local archaeological societies, willing volunteers and schools keen to embrace archaeology in their curricula. For archaeologists working in more economically impoverished regions, often but not exclusively in the Global South, community engagement can be both more challenging and more important. It is more challenging because communities struggle to make their daily living, lacking time for other concerns, and schools are under-resourced. At the same time, community engagement can be more important because communities can play a vital role in the protection of archaeological sites when either the legislative frameworks or capacities to implement those that exist are absent. Moreover, archaeologists from Europe and North America who work in the Global South are often following predecessors who operated in a colonial frame of mind. As such, they carry a social responsibility to make amends for what had been a frequent neglect of communities' interests and needs as their cultural heritage had been exploited for the benefit of imperial powers.

It hardly needs stating that neither using heritage to progress SDGs nor achieving positive forms of community engagement is easy to attain. Even without these tasks, archaeologists are already challenged by the demands of fieldwork, often undertaken in difficult environments and with limited budgets and time. They are rarely trained or experienced in either development-related disciplines or cultural heritage management, while there is no off-the-shelf formula for how to engage with communities and measure what benefits have been delivered. Communities are, after all, ill-defined and complex entities (Pyburn, 2011; Humphris and Bradshaw, 2017). As Gould and Burtenshaw (2019) noted, communities have individuals who *'may live together harmoniously, or not; they may have parallel economic interests, or not; and their identification with the heritage itself may range from profound to indifference'*.

Our motivation for preparing this volume arises from our personal experiences. All three of us have worked on projects in the Global South, Steven in Jordan, Mubariz in Pakistan and Iraq and Maria in Iran and Iraq. We have become concerned with how archaeologists from the Global North can operate in such regions without maintaining a form of colonial-like practice.

Steven's experience in Jordan during the 1990s and 2000s was that the British Institute in Amman, one of the British Academy-funded International Research Institutes that support UK interests overseas, imposed a colonial-like approach and experience upon visiting archaeologists, despite making every attempt not to do so: it had become embedded into the fabric of the institution.

That became of particular concern after having completed his 1996–2010 survey and excavation project in Faynan, a desert-like environment in southern Jordan occupied by a sparse and economically impoverished Bedouin community. Steven felt that having contributed to knowledge about the Neolithic and advanced his own career, he was leaving nothing for the local community without whom the fieldwork would have been impossible. He had an academic vested interest in this concern: irrigation-based farming was expanding in Faynan, threatening not only archaeological sites (discovered and unknown) but also Faynan's natural heritage that was a haven for birdlife (that was also becoming decimated throughout much of Jordan). How could the archaeological sites in Faynan be protected when archaeologists were no longer on the ground, people had to make their living, and the Department of Antiquities lacked the capacity to intervene?

His project co-director, Bill Finlayson, felt the same and was already engaged in community-focussed work at the nearby Neolithic site of Beidha and the DEEPSAL project that was exploring the deep past as a social asset for local communities in Jordan (Burtenshaw et al., 2019). Together with Mohammed Najjar, who had excavated in Faynan for many years, we embarked on a project to develop a local museum in Faynan, one that had been built by the Department of Antiquities but had remained an empty shell. Having installed some displays, Steven then attempted to make the museum, and the cultural heritage it represented, of greater value to the Faynan community, anticipating this would ultimately contribute towards the protection of archaeological sites. Following a request from the community, Steven then co-led a project to establish a handicraft business with the aim of generating income for local families from the tourists being attracted to Faynan.

Those projects were undertaken in collaboration with other academics from the UK and Jordan, with Jordanian not-for-profit organisations experienced in development projects, and the approval of the Jordanian Department of Antiquities. Working with relatively low budgets, across the COVID-19 pandemic years and as the Middle East went from one international crisis to the next, the projects were as challenging to undertake as the outcomes were problematic.

To document and make a critical analysis of the museum and handicraft projects, Steven invited Mubariz and Maria to co-author a journal article that placed the Faynan project into a wider context, feeling their knowledge, experiences and independent evaluation would balance his inevitably biased insider opinion.

Mubariz is a member of the International Association for the Mediterranean and the East (ISMEO) Italian Archaeological Mission in Pakistan and is part of the team led by Luca M. Olivieri (the author of the Archaeology, Community, Tourism (ACT) case study, Chapter 7 in this volume). When Mubariz joined the Archaeology, Community, Tourism – Field School (ACT) project in 2016, he participated not only in the excavations at the archaeological site of Barikot but also ensured, together with other team members, the implementation of the project's community engagement and tourism-related initiatives on the ground. As part of that project, he attended a seminar on 'Sustainable Archaeological Tourism and Role of Local Communities in Heritage Conservation' at the Swat Museum, which was jointly organised by the Sustainable Tourism Foundation Pakistan, the Archaeology Tourism Project and the Department of Archaeology in Museums with the support of the Suvastu Art and Culture Association and the Swat Archaeological Guides Association. Subsequent visits not only to Swat but also to Duhok in Iraqi Kurdistan (the sites under study by the Land of Nineveh Archaeological Project, Chapter 6 in this volume) further developed his interest in community archaeology-centred approaches and convinced him that without effective local community engagement involvement, archaeological research cannot be conducted in a sustainable, just and mutually beneficial way.

Maria's contribution draws from both her professional archaeological work and personal experiences. Traveling to numerous cultural and archaeological sites, particularly the UNESCO World Heritage sites of Borobudur and the Prambanan temple compounds in Indonesia, profoundly shaped her perspective. Observing the site management's efforts to attract visitors, increase public awareness, and conduct preservation activities highlighted for her the complexities inherent in sustainable tourism – particularly in economically vulnerable regions. It made her wonder about the complex series of challenges these sites face, their success in balancing preservation with tourism demands and how these efforts can be directed to benefit the local communities.

When the three of us, Steven, Mubariz and Maria, began working together to review and contextualise the Faynan Heritage projects, we required comparable projects to assess whether the problems faced in Faynan, the community engagement approaches adopted, and the outcomes achieved and missed, were typical or unusual. How can successes and mistakes be identified without reviewing what the alternative approaches might have been?

With some notable exceptions (e.g. Hodges, 2014), we struggled to find published reports of sufficiently similar archaeology-based heritage projects that described not just the positive outcomes, but also the 'warts-and-all' experience – neither authors nor journals like to publish declarations of weaknesses and mistakes, let alone failures. As Gould and Burtenshaw (2014, p.8) argued in their pioneering collection of articles on 'Archaeology and Economic Development', there is a:

> need for heritage practitioners to present their case studies, including both successes and failures, and to discuss candidly their motivation, goals and outcomes … a vision of best practice will only emerge when there is a far greater body of case study material to draw upon.

While published reports were sparse, we were nevertheless aware of archaeology-led projects that, in addition to pursuing research, were collaborating with local communities with dual aims of conserving cultural heritage and enhancing the economic standing and social well-being of those communities. We decided to write to the directors of such projects, asking if they would contribute a description of their projects to publish within a volume that contained comparable studies from elsewhere in the Global South. We stressed our wish for reports that not only described their projects but also reflected on their challenges, successes, failures and ongoing concerns. We explained there was no intention to make any evaluations, let alone judgements, but wished to compile a portfolio of studies that demonstrated the efforts that archaeologists were undertaking to deliver public good in the present while also researching the human past. Although being neither a textbook nor a handbook, we wanted a volume that would be of use to those beginning projects with the same intent – the type of book that Steven would have liked available prior to embarking on his own projects in Faynan.

Without exception, the project directors agreed, embracing the aims of the book. Indeed, ultimately, we had to be selective about which projects to include. We wished to have a volume that contained a diversity of archaeological site types, geographic locations, scales of funding and approaches to community engagement, while also having sufficient overlaps between the projects to explore common themes. To facilitate that, we chose to draw on those projects that involved, to a greater or lesser extent, a concern with supporting sustainable heritage-based tourism.

The resulting volume has fourteen case studies (Figure 1.1). We have arranged them by continent, seven from Asia, three from Africa and four from Latin America. They vary between single-author accounts to those produced by multinational project teams; some have more focus on the archaeology itself, while others swiftly pass over their sites to discuss the forms of community engagement adopted and contributions to SDGs; some projects have been completed, some are ongoing, while others describe their future plans.

The case studies begin with the projects in Jordan referred to above, those co-led by Steven Mithen in Faynan (Chapters 2 and 3) and by Bill Finlayson at Beidha (Chapter 4). In Chapter 5, we turn to one of the most significant developments in the Jordanian Heritage sector, the foundation of 'Sela for Training and Protection

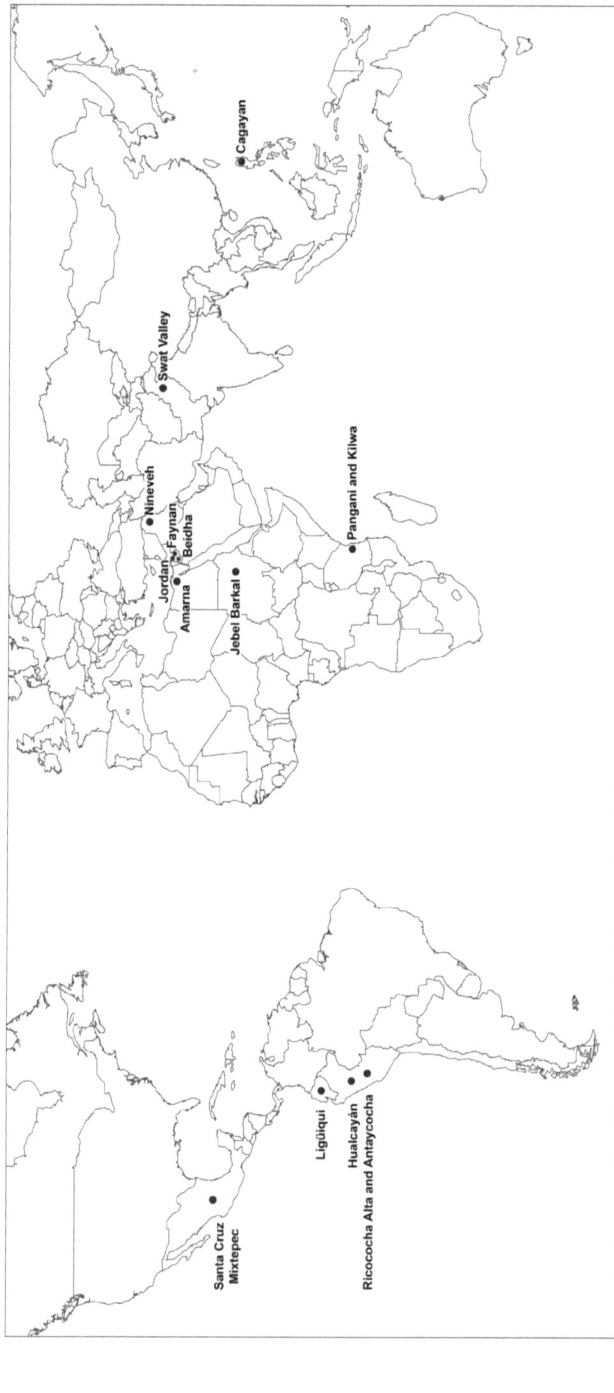

Figure 1.1 The archaeological sites and localities considered by case studies in this volume. Map courtesy of Cosmographics.

of Heritage' in 2015, a non-profit organisation based in Wadi Musa (Petra). Maria Elena Ronza, one of its founders and current president, reviews how this not-for-profit organisation provides protection and preservation of cultural heritage sites by promoting the active engagement of local communities and developing the local tourism sector.

The next two case studies are the largest and most developed in our volume. Both come from regions of recent conflict and have been substantially supported by Italian funding bodies and ALIPH, the International Alliance for the Protection of Heritage in Conflict Areas. They both work closely with the governmental bodies in their host countries, dealing with archaeological sites containing monumental art and architecture. Both are situated in the north of their respective countries in fertile and mountainous environments. In Chapter 6, Daniele Morandi Bonacossi and his colleagues describe the Land of Nineveh Project in the Kurdistan Region of Iraq that is developing an archaeological park for the remains of the Assyrian irrigation system of the 7th century BCE. A similar landscape scale approach is described in Chapter 7 in which Luca Maria Olivieri reviews projects undertaken by the Italian Archaeology Mission since 2011 in the Swat Valley of Pakistan. These have sought to study and restore archaeological sites damaged by conflict by using methods that enable economic development for the region.

Our final study from Asia comes from the Philippines. In Chapter 8, Caroline Lising and Grace Barretto-Tesoro describe the government-led desire to enhance heritage-based tourism in Cagayan Province and their interest in the neglected archaeological and historic sites of the Spanish Colonial era (15th–19th centuries). They describe the challenges arising from these sites having multiple stakeholders, the tensions between using churches for ongoing Catholic religious practice and tourism, and the need for training programmes in cultural resource management.

Our Africa section begins with Chapter 9 by Kate Spence and her colleagues, who describe their community-oriented work at Amarna, Egypt, the ancient city of Akhenaten, founded in 1340 BC. The site is threatened by creeping urbanisation and, although academically important, lacks the monumental architecture found elsewhere in Egypt that attracts mass tourism. In Chapter 11 Rebecca Bradshaw and her colleagues describe their work at a complex of archaeological sites surrounding Jebel Barkal in Sudan, a sandstone outcrop that was sacred to ancient Kushites and Egyptians (c. 1700 BCE to 350 CE). Their project seeks to undertake research on these sites, while protecting and conserving them in a manner that supports the interests of the local community who have long been suffering from armed conflict in the region. As from 2024, that conflict intensified leading to a humanitarian crisis causing Bradshaw et al. to ask (and answer) whether 'they care about ruins, artefacts and museums more than living people' and 'whether they should continue to work on heritage during wartime given such extreme human suffering'?

Our Africa section ends with Chapter 11 in which Paul Lane and his colleagues describe their work to promote a wider public understanding of Tanzania's built cultural heritage by involving communities in its documentation, protection and conservation. This derived from the premise that the built environment can be a crucial resource for sustainable socio-economic development and involved work with diverse stakeholders in two key areas of Tanzania, Pangani on the northern coast and Kilwa on the southern coast.

Moving continents, in Chapter 12 Alanna Cant starts our section on Latin America by describing a 2015–18 project to conserve and restore Santa Cruz Mixtepec a ruined 16th-century Dominican convent in Mexico. She questions how the 'local community' can be adequately defined, describes the specific challenges faced by religious heritage and explores the conflicting interests of stakeholders who nevertheless managed to collaborate in good faith and with considerable success. In Chapter 13, Juan Andrés-Jijón Porras and his colleagues describe the archaeological sites of Ligüiqui, Ecuador, that include underwater stone-built fishing corrals that testify to a continuity of marine exploitation for over 1500 years. The present-day community is facing many challenges, ranging from the impact of El Niño events to drug traffickers, leading Porras et al. to consider whether and how their archaeological heritage can be used as a means of sustainable development.

Our final two studies come from the Peruvian Andes. In Chapter 15, Rebecca Bria and Erick Casanova Vásquez describe their long-term engagement with the people of Hualcayán, a village of Andean farmers where extensive archaeological evidence documents 4000 years of Andean prehistory. They describe their approach as an 'archaeology of care' towards the past and present people of Hualcayán, recounting how relationships have evolved over 15 years of working together. One of the challenges facing the Hualcayán community arises from glacial retreat that places pressure on water supplies. That issue is the focus of Chapter 15, in which Kevin Lane and his colleagues consider past and present water management in the Peruvian Andes. They draw on a palaeoecological study from the Lima Region to illustrate the effectiveness of pre-Hispanic period stone-built dams that became abandoned during the colonial period. Lane et al. explain how recent attempts to recreate those dams by using concrete and steel have been ineffective and describe a more sustainable approach that draws on community knowledge, local materials and archaeological expertise, illustrated by the rebuilding of the Ricococha Alta dam and its impact on the local environment and economy.

As will be evident, while all 14 case studies are concerned with community engagement and economic development, they are diverse in their scale, types of archaeological sites, contexts in which they operate, challenges they face and approaches adopted. Each of them provides their own individual lessons for themselves as their work continues and for those who are embarking on similar projects. Despite having their unique characters and lessons, several common themes emerge from across our case studies. We identify and discuss these in Chapter 16 as our conclusion to this volume.

References

Baillie, B. and Sørensen, M.L.S. (2021). *African Heritage Challenges: Communities and Sustainable Development.* Singapore: Springer.

Burtenshaw, P., Finlayson, B., El-Abed, O., and Palmer, C. (2019). The DEEPSAL Project: Using the Past for Local Community Futures in Jordan, *Conservation and Management of Archaeological Sites*, *21*(2), pp. 69–91.

Cernea M. (2001). *Cultural Heritage and Development: A Framework for Action in the Middle East and North Africa*. Washington: World Bank Group.

Chirikure, S. (2021). Making Archaeology Relevant to Global Challenges: A Global South Perspective, *Antiquity*, *95*(382), pp. 1073–1077.

Giliberto, F. and Labadi, S. (2022). Harnessing Cultural Heritage for Sustainable Development: An Analysis of Three Internationally Funded Projects in MENA Countries, *International Journal of Heritage Studies*, *28*(2), pp. 133–146.

Gould, P.G. and Burtenshaw, P. (2014). Archaeology and Economic Development, *Public Archaeology*, *13*(1–3), pp. 3–9.

Gould, P.G. and Burtenshaw, P. (2019). 'Community Economic Development and Archaeology', in: C. Smith (ed.) *Encyclopedia of Global Archaeology*. Springer. DOI: https://doi.org/10.1007/978-3-319-51726-1_3385-1

Hampton, M.P. (2005). Heritage, Local Communities and Economic Development, *Annals of Tourism Research*, *32*(3), pp. 735–759.

Hodges, R. (2014). Escaping Enver Hoxha's Shadow: A Short History of the Butrint Foundation's Project in Albania, 1993–2012, *Public Archaeology*, *13*(1–3), pp. 11–29.

Humphris, J. and Bradshaw, R. (2017). Understanding 'the Community' Before Community Archaeology: A Case Study from Sudan, *Journal of Community Archaeology & Heritage*, *4*(3), pp. 203–217.

Labadi, S. (2022). *Rethinking Heritage for Sustainable Development. International Framework, Local Impacts*. London: UCL Press.

Labadi, S., Giliberto, F., Rosetti, I., Shebati, L., and Yildirim, E. (2021). *Heritage and the Sustainable Development Goals. Policy Guidance for Heritage and Development Actors*. Paris: ICOMOS.

Labadi, S. Giliberto, F. Taruvinga, P., and Jopela, A. (2020). *World Heritage and Sustainable Development in Africa: Implementing the 2015 Policy*. Midrand: African World Heritage Fund.

Madandola, M. and Boussaa, D. (2023). Cultural Heritage Tourism as a Catalyst for Sustainable Development; the Case of Old Oyo Town in Nigeria, *International Journal of Heritage Studies*, *29*(1–2), pp. 21–38.

Phillips, C. and Allen, H. (2010). *Bridging the Divide: Indigenous Communities and Archaeology into the 21st Century*. London: Routledge.

Pyburn, K.A. (2011). 'Engaged Archaeology: Whose Community? Which Public?', in: K. Okamura and A. Matsuda (eds.) *New Perspectives in Global Public Archaeology*. New York: Springer, pp. 29–41.

Waterton, E. and Watson, S. (2011). *Heritage and Community Engagement: Collaboration or Contestation?* London: Routledge.

Part I

Asia

2 The Faynan Heritage Project

Developing a Local Museum within a Rural Bedouin Community of Southern Jordan

Steven Mithen, Bill Finlayson, Fatima Al-Nammari, Carol Palmer, Gehan Selim, Jessica Jacobs, Elaine Jamieson and Nebras Maslamani

2.1 Introduction

Faynan is a region of c. 150 km^2 centred on Wadi Faynan in the Aqaba Governorate of Jordan, approximately midway between the southern end of the Dead Sea and the port of Aqaba. It has a visually impressive, academically important but unprotected archaeological landscape. Faynan is a region of considerable economic hardship, with high levels of unemployment and poverty amongst the local community. Might its cultural heritage be used to support sustainable economic development and might that development help conserve the cultural heritage? These were the questions asked in 2017 that led to the Faynan Heritage Project. That began with the development of a local museum to encourage and support ecotourism and developed to establish a small business to manufacture, market and sell local handicrafts (this volume, Chapter 3). In this chapter, we describe the landscape, population, economy and archaeology of Faynan, prior to describing the development of the museum, aiming to create an attractive resource for international visitors and a means to engage the Faynan community in its cultural heritage.

2.2 Faynan: Place, People and Economy

Faynan lies within and beyond the southern margin of the Dana Reserve, which is managed by the Royal Society for the Conservation of Nature (RSCN). Wadi Faynan is formed at c. 300 m asl by the confluence of Wadi Ghuwayr (sometimes spelt Ghwair) and Wadi Dana that cut their way down through the west-facing escarpment of the Jordanian plateau. Wadi Faynan flows westwards and merges into Wadi Fidan before reaching the Wadi Araba (Figure 2.1).

Faynan is a desert environment, with rainfall varying from a monthly mean of 0–0.1 mm between June and September to 17 mm between December and January. Flash floods can occur in the winter months. The landscape is a series of flat, broad valley bottoms and terraces, with scattered acacia trees and steppe vegetation. Riverine woodland grows in the vicinity of springs in Wadi Ghuwayr. Being

DOI: 10.4324/9781003491071-3

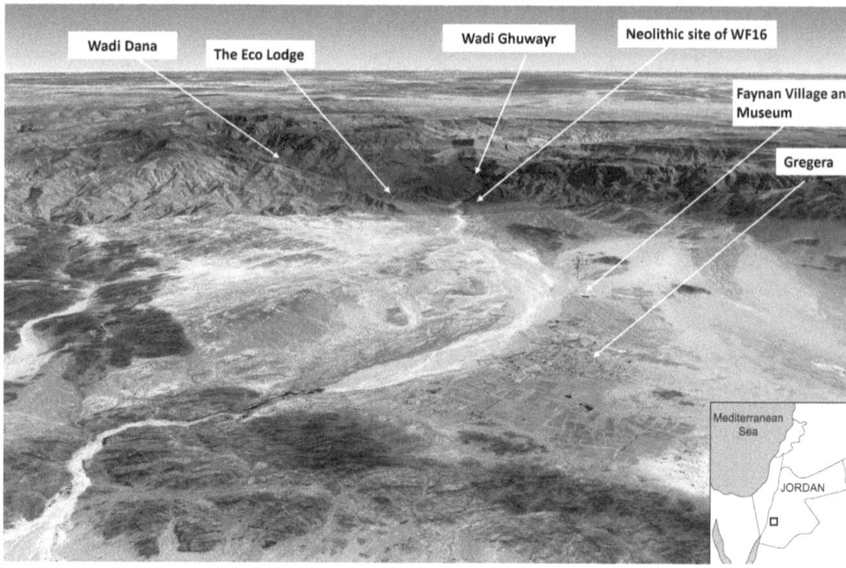

Figure 2.1 View of Faynan, looking east towards the Jordanian Plateau and marking key locations and features referred to in the text. © Steven Mithen.

located below the North–South Africa–Eurasia migratory flyway, Faynan has a particularly diverse birdlife with a mix of residents, seasonal migrants and visitors.

The Faynan population encompasses the communities within the Municipality of Qurayqira and Faynan. The village of Qurayqira (often spelt Greigra), located at the western extent of Wadi Faynan as it turns north to become Wadi Fidan, has c. 2000 inhabitants. The smaller village of Faynan is located to the east of Qurayqira within Wadi Faynan itself. Bedouin families also live in tents scattered across the landscape, most notably in the area around at the foot of the Wadi Dana. The main tribal groupings are the 'Ammarin, Sa'idiyyin, Rashaydah and 'Azazma.

The 'Ammarin and Sa'idiyyin are primarily represented in Qurayqira, which was established in 1976 as part of a government-led sedentarisation project. The Rashaydah were granted rights to establish their own village in the late 1990s, which became known as Faynan Village, also known as the Rashaydah village. The 'Azazma are a broad association, originating from Bedouin who were displaced from the Negev following the Arab–Israeli War of 1948. In the late 1990s, they were given the settlement of Gwebbe, north of Faynan, but many remain within tents, maintaining a traditional transhumance pattern of summer on the plateau and winter in Faynan.

In 2010, Faynan/Greigra was ranked 5th out of 17 governorates in Jordan for its level of poverty, with c. 50% of its population below the poverty line. By 2016, poverty levels had increased to 71.5%. Other than traditional goat herding, the economy is based on farming and tourism. Farming for watermelons and tomatoes

Figure 2.2 Faynan in March 2017, showing the development of irrigated tomato and melon fields over a Roman field system and desert. © Steven Mithen.

began in the vicinity of Qurayqira and expanded into the Wadi Faynan in 1990 by drawing water from springs in Wadi Ghuwayr for irrigation. Further expansion occurred with the establishment of cooperatives in 1997 and 2002 and the creation of boreholes. It has now transformed a large area of former wilderness-like desert into cultivation, much of it covered with plastic to conserve water (Figure 2.2). Watermelons ripen early and are transported to markets in Amman and beyond. Returns from the watermelons and tomatoes are unreliable because of competition from farming elsewhere in Jordan and fluctuations in the international market. The closure of export routes to Iraq and Syria during the Syrian Civil War (primarily 2011–2018), for instance, had an impact on Faynan with the result that many tomatoes were left rotting on the ground.

A 2011 report found that the expansion of farming had a limited impact on the availability of food for local consumption (Darmame et al., 2011). The income generated was restricted to only 10–12 families, and many villagers appeared unwilling to become involved. When interviewed, some proposed that the expansion of fields was as much to secure tribal rights to land and water as to alleviate poverty and enhance social welfare. Farming has, however, continued to expand since 2011, and the current validity of that report is unclear.

The expansion of farming threatens known archaeological sites and those yet to be discovered. Although this threat led to archaeological surveys in Faynan during the 1990s, the risk remains unabated. The natural heritage is also being destroyed as desert vegetation is converted to cultivated fields and then covered by plastic, notably a loss of habitats for birds. Their loss and reduction in the wilderness-like quality of Faynan threatens another source of income, that is from tourism.

2.3 Tourism in Jordan and Faynan

Tourism is central to the Jordanian economy.[1] In 2017, it contributed 19.4% of GDP having received 6.4 million international visitors, an increase of 33.5% since 2015. The tourist industry accounted for 18.1% of employment and 8.8% of investment into the country. One of the challenges of Jordanian tourism is its reliance on international visitors who make short stopovers at a limited number of sites, notably Petra, Jerash and Madaba, these having 465,000, 211,000 and 258,000 visitors, respectively, in 2017. Such tourists predominantly arrive by air or cruise ship, travel by coach, stay in internationally owned hotels and depart without making expenditure into local communities. Jordan's reliance on international visitors makes the tourism industry vulnerable to external factors. The number of tourists fell markedly during the Syrian Civil War, collapsed during the 2020–2022 COVID-19 pandemic, began to recover and was then reduced by the 2023-ongoing Gaza conflict.

These challenges were recognised in the 2017–2021 National Tourism Strategy, led by the then Minister of Tourism and Culture, Her Excellency Lina Annab. She noted the need to '*transfer part of the outbound revenues to domestic tourism by developing tourist products at various sites across the country to encourage citizens to visit*' and the need for '*local communities to create business opportunities and products with a higher demand*'.[2] The desirable shift is away from what Mowforth & Munt (2003) describe as conventional tourism to their 'New Tourism', more generally known as ecotourism.

The first significant tourism initiative in Faynan was the establishment of the Wilderness Lodge by the RSCN in 2008. The RSCN wished to integrate nature conservation in the adjacent Dana Nature Reserve with socio-economic development and financed the construction of this hotel at the base of Wadi Dana. This facility opened in 2008, coming under the ownership of Ecohotels, led by Nabil Tarazi (Figure 2.3), and now known as the Feynan Ecolodge.

The ethos of the Feynan Ecolodge is to provide high-quality experiences for visitors while having a low carbon footprint and supporting the local community. It provides vegetarian meals, makes limited use of electricity, which is generated by solar power, has its own wastewater processing system and undertakes considerable tree planting. The access road from Qurayqira has been left unmetalled so that visitors rely on transport by Bedouin trucks from a reception centre in the Rashaydah village. Visitors have opportunities for guided walks, to engage in a 'Bedouin experience', cooking and craft activities, and stargazing; many simply enjoy the solitude in the wilderness-like landscape. It was stated in 2011 that 70% of the Ecolodge income was used to support the local community via the supply chain for food and transport, and employment, benefitting 400 people locally.[3] These figures have been disputed by some members of the Bedouin community. During the last decade, the costs of staying at Feynan Ecolodge have steeply

Figure 2.3 (a) Location of the Feynan Ecolodge at the base of Wadi Dana, (b) the Feynan Ecolodge and (c) Bedouin trekking camp located at the base of Wadi Ghuwayr. © Steven Mithen.

increased, now (October 2024) exceeding those for a room in central London, and hence excluding anyone other than wealthy visitors. It has become unsuitable for independent travellers on limited budgets who could, nevertheless, contribute to the local economy via their expenditure.

Following the establishment of the Feynan Ecolodge, some members of the Bedouin community became aware of the tourist potential of Faynan and established their own trekking camps, notably at the base of Wadi Ghuwayr (Figure 2.3). These cater for those travelling on restricted budgets, often as members of trekking parties led by a local guide. The trekking camps enable tourist income to become more widely distributed, some arguing that the employment and supply-chain benefits of the Ecolodge are more limited than claimed.

Whatever facilities are available, the extent of tourism to Faynan will always be uncertain, because of competition from similar ventures elsewhere in Jordan and the constantly changing international situation. As such, Faynan needs to exploit all its assets to compete for tourists; while the Ecolodge and trekking camps promote the wilderness-like landscapes of Faynan, its cultural heritage has been neglected as a potential tourist attraction. A 2014 survey of visitors at the Ecolodge recorded their interests as 50% visiting for trekking, 30% for finding solitude, 25% for bird-watching and nature and 18% for the 'Ecolodge experience'. Only one person from 300 cited 'archaeology', despite the Ecolodge being situated on the doorstep of numerous ancient monuments, which are visually striking and historically significant (Burtenshaw, 2014).

2.4 The Archaeology of Faynan

Archaeological work has taken place in Faynan for over a century, identifying the region and the adjacent Wadi Fidan as key areas for the development of copper mining and its industrial development during the Roman and Byzantine periods. That involved the establishment of a major administrative centre, known as Phaino in the Classical texts, and now represented by the ruins of Khirbet Faynan, with associated mines, cemetery, aqueducts, mill and field system.

In 1988, Mohammad Najjar, Director of Archaeological Surveys and Excavations for the Jordanian Department of Antiquities, became a co-director of a project investigating the history of a copper mining project with Andreas Hauptman from Bochum Mining Museum, Germany (Hauptmann, 2007). Najjar began his own programme of archaeological research, starting with the excavation at the Neolithic site of Tell Wadi Faynan, where he found traces of the earliest copper working in the area (Najjar et al., 1990). In 1993, he investigated the remains of an earlier Neolithic site of Ghuwayr 1, located where the Wadi Faynan and Ghuwayr join, continuing his excavations in collaboration with Alan Simmons of the University of Nevada, Las Vegas (Simmons & Najjar, 2006).

During the early 1990s, the Roman/Byzantine cemetery was looted by local Bedouin in the hope of finding grave goods. The British Institute at Amman for Archaeology and Art History (BIAAH, later the Council for British Research in the Levant (CBRL)) and Yarmouk University undertook a rescue excavation of the largest cemetery site (WF3). As the threat from farming to the walls of the unrecorded Roman field system became evident, BIAAH/CBRL encouraged a programme of archaeological survey, rescue and research excavations, forming the flagship Wadi Faynan Project. The Khirbat Faynan was surveyed, and excavations were undertaken at a Bronze Age site (WF100) located close to the existing extent of farming.

Two larger projects followed: an extensive and detailed survey of the base of Wadi Faynan led by Graeme Barker and David Mattingly (University of Leicester, UK; Barker et al., 2007;) and survey of its side terraces, Wadi Ghuwayr and Wadi Dana by Bill Finlayson, who had become director of the CBRL in 1996, and Steven Mithen (University of Reading, UK; Finlayson & Mithen, 2007). Both projects were primarily funded by awards from the UK's Arts and Humanities Research Council (AHRC).

These projects expanded the list of the known archaeological sites to include scatters of Palaeolithic artefacts and an early Neolithic site predating Ghuwayr 1, designated as Wadi Faynan 16 (WF16). They greatly increased our knowledge about the gradual development of field systems, water management and copper mining from the Early Bronze Age through the Classical era and into the Ottoman Period. This work was further developed by a study focussing on water availability and usage throughout the past and into the present day in Faynan (Mithen & Black, 2011). Major programmes of excavation were undertaken by US teams in the

Figure 2.4 Excavations at Wadi Faynan 16 (WF16) in April 2010, with the pisé-constructed semi-subterranean structures. © Steven Mithen.

adjacent Wadi Fidan led by Tom Levy from the University of California (e.g. Levy et al., 2016). Between 2008 and 2010, Finlayson, Mithen and Najjar undertook an excavation at WF16, demonstrating this to be one of the largest and best-preserved early Neolithic sites in the southern Levant (Figure 2.4; Mithen et al., 2018).

A consequence of this fieldwork is that Faynan is now recognised as having an impressive range of archaeological sites. This includes both visually striking monuments situated in a beautiful landscape and academically important sequences documenting the transition from hunter-gathering to farming and the development of copper mining from its beginnings to an industrial scale in the Classical period.

Here, it must be noted that none of the Faynan tribes claims any connection with the archaeology of Faynan. They have all been relatively recent migrants into Faynan and have no sense of cultural ownership of the archaeological sites that for generations have been walked across while herding goats. Not only is there an absence of any connection or sense of ownership, but also a seeming lack of interest.

2.5 Unequal Beneficiaries

Archaeological fieldwork in Faynan provided economic benefits to the local community by the employment of workers and the purchase of food, fuel and other materials. In 2009, for instance, the 8-week excavation at WF16 spent £18,582 on wages for Bedouin workers, £1500 on fuel and £34,000 on accommodation and food. Equivalent amounts were spent in 2008 and 2010, with similar expenditures into the economy by other field projects since the 1980s. The flow of income largely ceased after 2010, since when there have been no large-scale field projects. Any archaeologist visiting Faynan today will likely be asked when they intend to start fieldwork and how many workers will be required.

The academics leading the fieldwork in Wadi Faynan and Wadi Fidan secured major, long-term benefits. By securing research grants and publishing articles and books, their reputations and careers flourished. The contrast with the absence of long-term benefits to the local community from the archaeological fieldwork is striking. This disparity, along with a concern about the ongoing threat to Faynan's cultural heritage, motivated the development of two projects that have since been encompassed together and labelled as the Faynan Heritage Project. The first, reported here, was the development of the Faynan Museum to support ecotourism; the second, reported in Chapter 3 of this volume, was to establish a small business utilising the cultural heritage of Faynan.

2.6 Developing Faynan Museum

The Department of Antiquities (DoA) of Jordan began to construct a local museum in 2000, encouraged by Mohammad Najjar. The museum was located on the edge of the Rashaydah village (Figure 2.5), with the building not completed until 2015 (after Mohammad Najjar had left the DoA to work as an independent consultant). The DoA invited Mohammad Najjar and Bill Finlayson, to construct displays in the museum, who in turn invited Steven Mithen to join them.

Rather than just seeking to procure display cabinets and artefacts about a range of key sites, it was decided to emphasise how archaeology is found throughout the

Figure 2.5 The Faynan Museum. © Steven Mithen.

Figure 2.6 The Faynan landscape model and wall displays. © Steven Mithen.

Faynan landscape. An award was secured by Mithen, Finlayson and Najjar from the UK AHRC funding body to construct a 3D model and associated wall displays.[4] The aim was to develop a resource that would be of interest to the local community, attract more tourists to Faynan and encourage day visitors to stay for longer. It was hoped that by increasing awareness about Faynan's archaeology within the community, greater protection would be afforded to monuments when making new fields, ponds and roads.

By using a digital elevation model of Faynan and an extensive photographic record, the model was constructed by a UK company, shipped to Jordan, transported to Faynan and assembled in the museum. Thirty archaeological sites were marked on the modelled landscape, with each site described on a surrounding wall display with text in English and Arabic. Sites were displayed in chronological order, taking the visitor from the Palaeolithic period to the Ottoman period (Figure 2.6). A full-colour guidebook to the archaeology of Faynan was also prepared and printed in Jordan.

The AHRC funds were limited, with none made available for cabinets to display artefacts; neither was funding forthcoming from the Jordanian Department of Antiquities (DoA). The DoA appointed Ali Hasaseen, who was both from Faynan and a graduate in archaeology, as the curator. The museum was opened on 4 March 2018 by Lina Annab, Minister for Culture and Tourism, with a speech from Monther Jamhawi, the Director General of the DoA and 200 hundred invited guests (Figure 2.7). Huge interest was shown in the landscape model.

Initial jubilation from the opening of the museum soon turned to frustration as what had been assumed to be the next steps of the project failed to materialise. Ali Hasaseen was only intermittently at the museum, and consequently,

Figure 2.7 Opening of Faynan Museum by Her Excellency Lina Annab, Minister for Culture and Tourism and Professor Steven Mithen. © Steven Mithen.

it remained locked with no information about how a visitor might gain access. More seriously, the location of the museum was soon identified as being problematic. Quite why it had been placed at the edge of the Rashaydah village was unclear. While this gave the Rashaydah tribe a sense of ownership, it deterred those from other tribes from visiting the museum, despite Ali Hasaseen being 'Ammarin. This impeded the idea that the museum could be a facility for the whole community.

Because the museum had a large parking area and spare office space, it provided an ideal venue for the Faynan Ecolodge reception, which was currently located a kilometre away at the western end of the village. That is where guests arrive in their own vehicles or by taxis and coaches, have a short rest and are then taken by Bedouin in their own trucks along the 8 km of un-metalled track to the Ecolodge. By moving the reception to the museum, the guests would have been able to visit the museum on arrival and gain an appreciation of the cultural heritage of Faynan before their transport to the Ecolodge. Ideally, a small shop would have been established to sell them the guidebook and ultimately souvenirs of their visit as income for the local community.

The Rashaydah were insistent that they would not move the reception centre to the museum. They were currently paid a rental for the existing reception by the Ecolodge and feared the loss of that revenue. Nabil Tazari, founder of Ecohotels who runs the Ecolodge, offered to continue the existing payment, but to no avail.

Repeated requests to the DoA to intervene had no impact. The DoA also refused to consider opening a shop at the museum because it was against their museum policy. Such intransigence from both the Rashydah and the DoA left the museum and reception centre for the Ecolodge unconnected. This failed to capitalise on the benefits to all parties that would have arisen had they been co-located.

2.7 Our Past, Our Future, All Together in Faynan

An opportunity to either break this impasse or work around it arose in 2018 when the UK Newton Fund was established. This was a UK fund to support collaborative research projects with countries in Africa, Asia and Latin America to support economic development and social welfare and tackle global challenges. Several country-specific schemes were developed, with that for Jordan designated as the Newton-Khalidi Fund. A significant proportion of the Newton-Khalidi Fund was to be allocated for arts and humanities-based projects, in contrast to the science focus elsewhere. As with all partner countries, Jordan was supposed to provide matching funds, but these did not materialise, although some contributions in kind, such as staff time, were provided.

Following a partner-building workshop event in Amman and a successful application to the Newton-Khalidi Fund, Steven Mithen and Fatima Al Nammari (University of Petra) became co-PIs of a multi-disciplinary project entitled 'Our Past, Our Future, All Together in Faynan', known as OPOF. This aimed to enhance the attractiveness of the museum to tourists and engage the Faynan community with their cultural heritage by developing a sense of ownership of the museum and their knowledge of the archaeology within the Faynan landscape. We anticipated this would facilitate and enhance their own initiatives, such as establishing trekking camps and acting as walking guides for visitors.

In addition to the Universities of Reading and Petra, the project CO-Is and institutional partners were:

- Carol Palmer, Director of the CBRL in Amman (successor to Bill Finlayson and BIAAH), with expertise in Bedouin lifestyles.
- Fatma Marii, an archaeologist with expertise in artefact conservation from the University of Jordan.
- Jessica Jacobs, a human geographer from Queen Mary College, London, with expertise in heritage, tourism and community engagement.
- Gehan Selim, an architect from the University of Leeds with expertise in architecture and creating spaces for education and heritage.

Nebras Maslamani, a recent graduate from Jordan who wished to build a career in sustainable development, was appointed as a project assistant, based at the CBRL in Amman. Project support also came from Arwa Badran (Durham University) with expertise in engaging with Jordanian schools and Elaine Jamieson and Sarah Lambert-Gates (University of Reading) with expertise in cultural heritage-related design and graphics. During the development of the project, two not-for-profit Jordanian organisations also became partners:

- Future Pioneers for Empowering Communities,[5] with Obayda Hummash as CEO.
- Sela,[6] led by Maria Elena Ronza, which has the aim of engaging Jordanians in the protection of their cultural heritage (see this volume, Chapter 5).

Five work packages were proposed, modified following meetings with members of the local Faynan community and approved by Ms Samia Khouri, Director of Museums and Awareness in the Jordan Department of Antiquities. The packages evolved as the project developed and were all implemented with varying levels of success.

2.7.1 Extending the Museum Displays to Cover Recent History

In hindsight, an error when developing the Faynan Museum had been to end the wall displays in the Ottoman period. That created a dislocation of the present from the past and provided no information to visitors about one of the most interesting aspects of present-day Faynan, the traditional Bedouin way of life and intangible heritage. As such, Carol Palmer provided text for wall displays to cover the history of Faynan during the last 100 years, involving the foundation of Jordan, and the Bedouin tent, food and culture. These were supplemented by an assortment of traditional Bedouin items, representing various items including baskets, clothing, jewellery and musical instruments.

2.7.2 Craft Workshops and Wall Hanging

Maps are by their nature a reduction, with most being made by people on the outside looking in (Jacobs 2020). That was certainly the case for the landscape model and associated wall displays within the museum, imposed by a Western academic perspective on the modern-day landscape and Faynan's past. That inevitably meant that important memories, sites and processes that make up the lived realities of the Faynan residents were excluded, these displaced by external academic values. To address this deficit, OPOF decided to support the production of a wall hanging for the museum by the local community to capture their perspective on the landscape and its history, expressed through image, colour and texture.

Between 2019 and 2021, Jessica Jacobs and her colleagues Emily O'Mara and Andrea Morreau carried out a series of workshops with women from different tribes based in Greigra and Wadi Faynan to explore and share local craft skills. They found many older women with considerable knowledge and ability in handicrafts; most of the younger generation, however, did not see the value of such skills. Their workshops aimed to increase the perceived value of those skills by linking them to the production of a beautiful wall hanging, for which all women would have a sense of ownership.

Fifteen women from different tribes worked collaboratively with Jessica, Emily and Andrea to produce a handcrafted community map forged from a Bedouin tent that told their stories of place. The 'making space' environment was convivial and welcoming, enabling participants to share their crafting knowledge and stories of place, the details of their daily, monthly and annual routines, their pasts and their hopes for the future. Older women acted as trainers, passing on their traditional craft skills of weaving, embroidery and beading to younger participants. Their stories were sewn, felted and woven into five different storyboard panels, each panel depicting a different time of the day in their lives, with a longer timeline of the lunar year and months at the top and bottom of the panels.

The storyboard panels were made using a wet felting method as background. Scenes were then added using dry felting techniques and then embellished with embroidery, adding shells and other materials such as sequins. Loosely equivalent to the five prayer times and read from right to left, each panel told a timeless 'day in the life of' the village with scenes telling stories of sheep and goat grazing, the flood, harsh summers and wedding feasts at night.

The emphasis on the place was followed through by the use of local materials where possible and engaging with local practices of woolgathering that served to emphasise the cultural heritage of pastoralism and handicraft skills and heritage expressed in this way as a valuable expression of their identity. Bedouin women of all ages could see that representing stories about their daily lives in this way created an important symbol of cultural heritage, in fact a series of symbols that they could retain ownership of and be developed into a brand identity that could also lead to direct and indirect economic and social benefit.

Intent on rendering the makers more visible but careful to avoid direct representation, the team created icons of the women as 'caryatids' to highlight the Bedouin tent as a piece of architecture and to represent the way in which Bedouin women 'hold up' their communities (Figure 2.8). However, the Bedouin women caryatids were not just emblems of a Bedouin past; one carried a modern handbag, while a second had a child sitting in front of them on a laptop: these women didn't just represent the past, but they represented the future of Wadi Faynan. The caryatid-like representations of Bedouin women were used to link the storyboard panels when they were attached to the side of a Bedouin tent to create the final wall hanging. This is to be displayed in the Faynan Museum, installed above the landscape model of Faynan.

2.7.3 Working with Schools and Children

While there is a flourishing cultural heritage industry in the UK seeking to engage children in the past, contributing to their education and entertainment, this is effectively absent in Jordan. With the likely ongoing expansion of farming and associated infrastructure, future protection of Faynan's cultural heritage is partly dependent upon ensuring that children grow up with an understanding of its value and potential for generating income via ecotourism and hence have a vested interest in its protection.

Figure 2.8 The wall hanging. © Steven Mithen.

In 2018, there were six schools in Faynan: four in Qurayqira, one in the Rashaydah village and one serving communities near to the Feynan Ecolodge. Four were co-essential (primary) schools and two secondary schools, both in Qurayqira. In the academic year 2017/2018, there were 938 students attending the schools for the summer term compared with 1,048 students attending the schools in winter, with the gender ratio slightly favouring girls (53% to 47%).

Carol Palmer, supported by her CBRL colleague Firas Bqa'in, led a consultation programme with teachers from each of the schools regarding how OPOF might support their educational activities. This identified an absence of basic facilities and a need for teaching training, both going beyond the remit and the resources of OPOF to resolve. A series of visits to the museum were organised for children to see the landscape model, wall displays and artefacts (Figure 2.9). Sela arranged activities for school children including piecing together fragments of ceramic vessels, pottery production and making mosaics from tiles for 1st–6th graders, with copper jewellery production for 7th to 10th graders (Figure 2.10). These drew

Figure 2.9 School children visiting the museum and exploring the landscape model. © Steven Mithen.

on archaeological materials that the children would come across on the ground in Faynan and provided a context for teaching about Faynan's past.

Arwa Badran was commissioned to produce a cultural heritage resource book for both teachers and children. This told the story of Faynan, focussing on the two key developments represented by its archaeology, the origin of farming and copper mining. Sela also used Faynan and its past as the subject for a children's cartoon book in a series based around two characters, Bassam and Karam – the first books in Jordan about cultural heritage designed for children. 'Bassam and Karam in Faynan' was written by Rash Dababneh, who undertook book readings at Faynan Schools and gave signed copies to pupils and teachers. Funding was later secured to convert the book and a second story into animations, which were launched at the Museum of Jordan in March 2023.

2.7.4 Faynan Heritage Walking Trail

The Faynan landscape provides an exceptional blend of natural and cultural heritage, ideal for trekking, birdwatching and visiting ancient sites – all appealing to tourists. Trails are important tools for heritage re-imagination, landscape appreciation and understanding of the past (Storey, 2010). In 2018, the Ecolodge was providing guided walks for its visitors, but its local guides had little knowledge

Figure 2.10 Sela undertaking cultural heritage craft activities with children. © Sela.

about the archaeological sites they were walking over. Similarly, there was nothing to inform independent visitors about the striking archaeological monuments and ruins other than the museum displays, but access to the museum was often impossible. With the agreement of the DoA and RSCN, several information boards were designed to be placed discreetly in the landscape, seeking to be as unobtrusive as possible, while both informative and sufficiently robust to survive the intense summer heat, winter cold, rain and likely vandalism. The boards were designed by Steven Mithen and Elaine Jamieson, with eight positioned close to key monuments and arranged in a sequence to provide a walking trail through major events in Faynan's past. Five more were located in more distant locations, beyond easy walking distance but to provide a guided trek to visit major archaeological sites. The boards were positioned on stone-built platforms, made using local materials (Figure 2.11). Additional boards providing a plan of the heritage trails and treks were made for positioning at the museum, Ecolodge and Bedouin trekking camp.

Sela was commissioned to install the information boards, using this as an opportunity to employ local workmen who were taught relevant skills. They were also employed to remove large metal posts concreted into the ground that surrounded the Neolithic site of Ghuwayr. These had once supported a wire netting fence, the wire netting having long been reused elsewhere, leaving the remaining posts as an eyesore.

Soon after erection, several of the information boards and platforms were intentionally damaged, having been struck with either stone or metal objects, and in

Figure 2.11 Board located close to the Roman aqueduct and the Byzantine cemetery on the Faynan Heritage trail. © Steven Mithen.

some cases destroyed with considerable effort. The culprits were older children, with most damage done to the boards close to their walking routes to and from school; those at a further distance were unaffected. This is an unfortunate problem throughout Jordan, one that provided a further incentive for OPOF's work with the schools. Having anticipated this problem, robust materials had been used for the boards, and hence, damage was often limited to scratches and dents rather than fatal fractures as occurs elsewhere in Jordan when, say, ceramic materials are used.

Sela undertook four rounds of maintenance and repairs between June 2022 and 2023. The extent of vandalism and damage decreased through that period, hopefully because of Sela's children-centred community activities in Faynan.

2.7.5 Re-imagining the Museum as a Cultural Hub

Museums around the world have evolved from being mere buildings in which antiquities from the past are housed into centres for cultural activities that contribute towards social cohesion and the well-being of local communities. Unfortunately, Faynan Museum did not offer a rewarding experience and was largely neglected. Might the Faynan Museum also play such a role for the Faynan Community? Even with the landscape model and wall displays, it had unused rooms that could be used for meetings, social events and children's activities. A first step would be to develop a social space for adults and a play space for children in the immediate vicinity of the museum, which would also make it an attractive location for the reception centre for the Ecolodge, creating the much-needed community cultural heritage hub for Faynan. Developing the museum in this manner would involve funds far beyond those available to OPOF. It was, however, within the project's capacity to undertake a design process to establish a vision for the Faynan Museum, one that future funders might support.

This design process was undertaken in two stages. The first was led by Fatima Al-Nammari and Yasser Abu Hashem, from the Department of Architecture, at Petra University. Ten undergraduate architecture students visited Faynan for workshops with members of the local community, discussing ideas for how the museum might be developed into a community hub. Three groups of Faynan participants were identified and invited: women, men and children. Men discussed the economic role of the museum mainly, while women and children imagined a new space for interaction. Children drew sketches of their favourite games and what they would like to have inside the museum and within its surrounding space. Ali Hasaseen, the museum curator, supported the workshop and represented the views expressed by the Department of Antiquities. Back at their university, the students were then tasked with producing a design for the outside space of the museum that would

Figure 2.12 Faynan Museum reimagined as a cultural hub for the Faynan community. © Gehan Selim.

help to transform it into a community hub, meeting the wishes of the community, and the needs of the Department of Antiquities, and that would appeal to visitors to Faynan. Another visit was made to Faynan to discuss the final designs with the community.

The students' design ideas provided the foundation for the second stage of this project led by Gehan Selim at the University of Leeds. She drew upon these to produce a final design for the outdoor space (Figure 2.12). This proposed using traditional building materials and methods to create a series of play activities for children around a central gathering area for meetings, talks and socialising and a space of market stalls for selling craft products and produce. The play activities consisted of a maze, climbing wall, hopscotch, face-in-the-hole game and sandboxes.

2.8 Reflections

The OPOF project was challenging, not least because its timeframe encompassed the COVID-19 pandemic that delayed and disrupted each of the work packages. A fundamental hurdle had been the location of the museum in the Rashaydah village that constrained full community participation and the intransigence of the Rashaydah to relocate their Ecolodge reception centre to be adjacent to the museum. Further unexpected issues arose, notably vandalism of the walking trail boards and leaks in the roof of the museum that delayed the installation of the wall hanging. On 19 March 2022, Her Excellency Mrs Haifa Najjar, the Minister of Culture for Jordan and Mrs Bridgit Brind OBE, the British Ambassador of Jordan, visited the Faynan Museum to view the work undertaken by the OPOF project. A presentation was made about the vision for the museum, leading to an invitation to meet with the Minister of Tourism and Antiquities in Amman for a more formal presentation. Seeking to measure the success or otherwise of OPOF is difficult. Funding did not allow any ongoing monitoring of the manner and extent to which the museum was used by the local community and by visitors to Faynan. Nor can the long-term impacts of the workshops, schoolbooks and children's activities be confidently assessed. Moreover, soon after the Israel–Gaza conflict led to a further collapse of tourism in southern Jordan, just as it was recovering post-COVID-19.

Along with the other Newton-Khalidi projects, OPOF was invited to submit a case for the 2020 Newton Prize for Jordan. That required a report describing OPOF's aims, activities and outcomes that were assessed by an international expert committee. It gained approval, winning the Newton Prize for Jordan. That provided funds for a follow-on project to develop a small business owned, managed and delivered by Bedouin women from Faynan to produce handicrafts for sale to tourists. That project is described in the following chapter of this volume.

Acknowledgements

The authors are grateful to the UK AHRC and the Newton Fund for funding the museum development projects in Faynan. Staff at the University of Reading administered the awards, while key contributions were made by members from all the institutions represented by the project members. Firas Bqa'in of the CBRL was especially supportive of the project. We are especially grateful to the local community in Faynan who supported our work – we hope you are benefitting from the outcome.

Notes

1 The following figures are taken from *Travel and Tourism Economic Impact Report 2017 for Jordan.*
2 Lina Annab, quoted by Oxford Business Group, 15 February 2017, https://oxfordbusinessgroup.com/news/positive-momentum-established-jordan%E2%80%99s-tourism-industry (accessed September 2017), and quoted in *The Jordan Times*, 24 July 2017.
3 These figures are provided to Steven Mithen by Nabil Tarazi.
4 AHRC AH/P005594/1, 'Discovering Faynan Heritage' project.
5 Future Pioneers for Empowering Communities (FPEC) is a Jordan-based non-profit organisation specialised in community development with a focus on women, youth and marginalised groups. Full details can be found on its webpage: https://future-pioneers.org/
6 Sela is a non-profit company founded in 2015 (Reg.# 712/2015) by five members of the communities living in the surroundings of the World Heritage site of Petra with the aim of actively engaging Jordanians in the protection of their cultural heritage. Full details can be found on Sela's webpage: https://selajo.org/

References

Barker, G., Gilbertson, D. and Mattingly, D. (2007). *Archaeology and Desertification: The Wadi Faynan Landscape Survey, Southern Jordan.* Oxford: CBRL/Oxbow Books.

Burtenshaw, P. (2014). *The Economic Capital of Archaeology: Measurements and Management.* PhD dissertation, University College London.

Darmame, K., Nortcliff, S. and Potter, R. (2011). 'The Role of Water and Land Management Policies in Contemporary Socio-Economic Development in Wadi Faynan', in: S. Mithen and E. Black (eds.) *Water, Life and Civilisation.* Cambridge: Cambridge University Press, pp. 442–453.

Finlayson, B. and Mithen, S. (2007). *The Early Prehistory of Wadi Faynan.* Oxford: CBRL/Oxbow Books.

Hauptmann, A. (2007). *The Archaeometallurgy of Copper. Evidence from Faynan, Jordan.* Berlin: Springer-Verlag.

Jacobs, J. (2020). Fabricating Herstory: Using Embroidery to Map Tribal Borders in South Sinai, *Journal of Arts and Communities,* 11, pp. 109–129.

Levy, T., Najjar, M. and Ben-Yosef, E. (2016). *New Insights into the Iron Age Archaeology of Edom, Southern Jordan.* Los Angeles: Cotsen Institute of Archaeology Press.

Mithen, S.J. and Black, E. (2011). *Water, Life and Civilisation: Climate, Environment and Society in the Jordan Valley.* Cambridge: Cambridge University Press.

Mithen, S.J., Finlayson, B., Maricevic, D., Smith, S., Jenkins, E. and Najjar, M. (2018). *WF16: Excavations at an Early Neolithic Settlement in Southern Jordan.* London: CBRL. https://www.jstor.org/stable/j.ctv37vwf8b

Mowforth, M. and Munt, I. (2003). *Tourism and Sustainability: Development and New Tourism in the Third World*. New York: Routledge.

Najjar, M., Abu Dayya, A. and Suleimann, E. (1990). Tell Wadi Feinan: The First Pottery Neolithic Tell in the South of Jordan, *Annual of the Department of Antiquities of Jordan*, 34, pp. 27–56.

Simmons, A.H. and Najjar, M. (2006). Ghwair I: A Small Complex Neolithic Community in Southern Jordan, *Journal of Field Archaeology*, 31, pp. 77–95.

Storey, D. (2010). 'Using the Past: Heritage and Re-Imagining Rural Places', in: *Geographical Perspectives on Sustainable Rural Change*. Brandon, Manitoba: Rural Development Institute, Brandon University, pp. 374–383.

3 Developing a Heritage-Themed Bedouin Handicraft Business in Faynan, Jordan

Success, Failure and Reflections

Steven Mithen, Obyda Hummash, Fatima Al-Nammari, Nebras Maslamani and Yusra Al-Jabari

3.1 Introduction

Faynan is an economically impoverished region in southern Jordan, with an outstanding cultural and natural heritage (see this volume, Chapter 2, for a full description). Following extensive archaeological research in Faynan primarily by UK-led teams, the Faynan Heritage Project was established in 2017 to use the cultural heritage of Faynan to support sustainable development via ecotourism. Its first stage involved developing a local museum to attract visitors and enhance the local community's engagement with Faynan's cultural heritage (this volume, Chapter 2). That initiative was awarded the Newton Prize for Jordan in 2020. That enabled a second stage for the Faynan Heritage Project, which sought to establish a small business to manufacture and sell handicrafts inspired by Faynan's cultural heritage to tourists. This chapter describes the opportunity, process, challenges and, regrettably, the ultimate failure of establishing that business, seeking to identify lessons learned that may be of value to equivalent projects elsewhere.

3.2 The Opportunity for a Handicraft Business in Faynan

During the museum-related workshops conducted with the Faynan community in the spring of 2018 (this volume, Chapter 2), local women approached Nebras Maslamani, the project manager, and Fatima Al-Nammari, the project Co-PI, encouraged by the fact that they were both women with leadership positions. They talked about their own handicraft products and the difficulty of accessing the tourism market. Their products included embroidery, beads and sewing, but were of relatively low quality to those on sale elsewhere in Jordan and, unbeknown to the women, had limited appeal to international tourists. At a community meeting in October 2018, local representatives asked for a project that would provide women with the skills to make products suitable for sale to tourists. This reflected a desire

DOI: 10.4324/9781003491071-4

to gain direct financial benefit from any tourists who might be attracted to Faynan by its Ecolodge, trekking camps and museum.

The market for gifts and mementoes is substantial in Jordan. Most of the products on sale are of a generic 'Bedouin' nature – beads, bags and trinkets – with a substantial number being cheap imitations imported from India and China. In 2018, the only point of sale in Faynan was a small shop in the Ecolodge, owned by the Royal Society for the Conservation of Nature. That sold a limited range of items, none of which were made in, or related to, Faynan. As such, the development of locally made products inspired by the cultural heritage and landscapes of Faynan appeared a promising initiative to meet the needs of the local community while also promoting interest in the archaeology of the region. This was precisely the request made in 2017 by the Jordanian Minister of Tourism and Antiquities to develop '*tourist products at various sites across the country*' and for '*local communities to create business opportunities and products with a higher demand*'.[1] More broadly, developing a handicraft business in Faynan would support the UN Sustainable Development Goal (SDG) numbers: 1 (no poverty), 5 (gender equality) and 8 (decent work and economic growth).

Mithen, Al-Nammari and Masalmani recognised that seeking to implement that vision in Faynan was high-risk and demanding. There was no evident tradition of quality craft activities within the local community, and the remote, rural, tribally organised and socially conservative setting of Faynan did not lend itself to establishing a self-sustaining business. Despite being explicit about the challenges, in 2019 the UK Arts and Humanities Research Council funded a project entitled 'Archaeology into Business'.[2] From November 2020, that project was subsumed into the 'Present Empowers the Past' (PEP) project using the Newton Prize, awarded for the Faynan Museum development (this volume, Chapter 2).[3] This new project sitting within the Faynan Heritage aegis was co-led by Steven Mithen (University of Reading) and Fatima Al-Nammari (University of Petra), with Nebras Maslamani as project manager. The project's aim was to establish a small business owned, managed and delivered by Bedouin women in Faynan for the production and sale of locally inspired handicrafts for their own and direct financial gain.

PEP's first step was to appoint an experienced project partner, for which the not-for-profit Future Pioneers for Empowering Communities (FPEC) was selected.[4] FPEC, led by Obyda Hummash, had an excellent track record of supporting the socio-economic development of disadvantaged communities in Jordan, helping them cope with climatic change, and enhancing their civic engagement. FPEC had already led a successful project to empower the local women and youth of the Souf and Sakeb villages in northern Jordan to initiate and sustain eco-tourism initiatives. That experience complemented that of Fatima Al-Nammari who had previously developed income-generating projects with women in Palestinian refugee camps in Jordan. As such, a wealth of expertise and experience was available, supported by Nebras Maslamani who had become familiar with the Faynan community. She worked within FPEC until 2022, after which her role was passed to FPEC employee Yusra Al-Jabari.

Following a series of meetings, PEP gained support from the Ministry of Tourism and Antiquities (MoTA), the Ministry of Culture, the Municipality of

Qurayqira and Faynan and the Wadi Araba Development Company. None of these, however, contributed funding while all provided their own bureaucratic hurdles for the project to surmount.

The key tasks for the PEP project were:

1. Establish a 'Faynan Heritage Women's Cultural Association' (FHWCA), securing its legal status and developing a small business enterprise, to be known as Faynan Heritage Home.
2. Identify a series of products to be manufactured by the FHWCA that would combine existing and new craft skills of the women, utilise designs based on Neolithic artefacts from Faynan and be attractive to ecotourists for purchase as mementos of their visit.
3. Provide training in craft and business skills, including marketing, accountancy and management.
4. Provide infrastructure, consisting of a business centre, an equipped workshop for manufacturing the products, a display area and selling point.
5. Identify points of sale for the products elsewhere in Faynan and throughout Jordan, notably in Aqaba and Amman.
6. Provide financial support and mentoring to the FHWCA, aiming for Faynan Heritage Home to become a sustainable business by the end of the funding period in spring of 2023.

Figure 3.1 Launch of Faynan Heritage Home by Mrs Haifaa Al-Najjar, the Minister of Culture for Jordan (centre left, pointing), and Mrs Bridget Brind, the British Ambassador (centre right, facing), 10 March 2022. © Steven Mithen.

The project started in October 2019, achieving the establishment of the FHWCA by the spring of 2021. Its business arm, Faynan Heritage Home, was launched on 10 March 2022 with an event attended by Mrs Haifaa Al-Najjar, the Minister of Culture for Jordan, and Mrs Bridget Brind, the British Ambassador (Figure 3.1). They and other attendees were able to visit the workshop, view and purchase a series of Faynan Heritage products and learn about future developments. The event was covered by the most important media agencies in Jordan, notably Ro'ya and Mamlaka, and reported on national TV.

The AHRC and Newton Prize funding streams ended in March 2023, after which support to the FHWCA had to be gradually reduced and ultimately ceased in December 2023. By March 2024, activity at Faynan Heritage Home had stalled, not having achieved a minimal level of sustainability – although a low number of sales were continuing at outlets it had secured. Whether the business can be reactivated by a new input of funding, support and personal is unclear. The following reflects on the steps taken for each of the above tasks, the challenges faced, successes, failures and lessons learned.

3.3 Establishing the Faynan Heritage Women's Cultural Association (FHWCA)

The creation of a new organisation was deliberated in depth because the Faynan community already had varied local organisations, although their purpose, membership and governance were obscure. Each was known to be affiliated with just one of the local tribes, creating conflicting power structures while limited resources caused an environment of competition. Attempts to merge organisations had failed because the deeply held tribal culture prevented collaboration. Nevertheless, because the idea for a new handicraft business was initiated by women, the creation of a new organisation was encouraged by the whole community.

Towards the end of 2019, women from Faynan were invited to join the FHWCA. They were asked to express their interest, interviewed and assessed according to their existing skills, experience and financial need. Emphasis was placed on selecting women from each of the five tribes to build social cohesion and share existing expertise and knowledge. Twenty-five women were appointed from 36 applicants, with seven placed into administrative positions to form the FHWCA leadership group.

A series of training sessions ensued, covering skills ranging from business management and accounting to pottery and weaving. A community development approach was adopted, ensuring that the women themselves were involved in the planning and implementation of all endeavours. Consequently, all appointed women received training in either craft skills and/or business and developed a role within the organisation. Legal status for the FHWCA was applied for and awarded in January 2021.

Both the idea and implementation of the project were received with mixed emotions by the local men. While some appreciated the idea of having a space that could aid women in making and marketing products, others, especially heads of local organisations, decried this new association and felt threatened. Nebras Maslamani, the project manager, visited all such organisations offering cooperation and opening the doors for joint work that could aid the local women. The stipulation was made, however, that women from all tribes must have an opportunity to be included – this following the ethos of the PEP project and reflecting the wish of the women themselves. That, however, was effectively rejected, with the men imposing a view that tribes should work by themselves. Although some tribes did not mind collaborating, they would still identify a specific tribe that they rejected.

The women forming the FHWCA's leadership group were devoted to the project, hardworking and talented, similarly for several of its members. Their achievements were substantial: acquiring skills, developing the business, generating products and achieving sales. With its external funding, the project continued to provide training to enhance their skills, develop new products and find market outlets. Raw materials were provided for women to start the production of marketable items; innovative product designs were developed drawing on archaeological finds from the Neolithic site of Wadi Faynan 16. Within a year, the women were thrilled to see the corpus of objects they were now producing that were both unique in design and created locally; they were even more pleased as sales began with the income distributed between the FHWCA members.

Just as the project was gaining success, and perhaps because of this, several rumours began to be spread about the project that had a negative impact on its progress. Nebras Maslamani and Fatima Al-Nammari were often asked about the level of project funding (coming from the UK), claiming that it should all be given as cash to the women. The promotion and advertising of the size of the Newton Prize funding online by UK funding bodies was unhelpful. This was seen by members of the Faynan community and interpreted as meaning that the PEP project had vastly more funds available than was the case. They requested that all the funding should be given directly as cash to the women.

Some direct payments were being made, following local norms for training and travel time, but the amounts were soon declared as unacceptable. Some men accused the project of benefiting from local women or attempting to disrupt the local order of things – despite the request for the project having come directly from the women themselves. Under such pressure, maintaining commitment from members of the FHWCA became difficult. Some withdrew, giving reasons of sickness and family commitments. Although unspoken, in some cases, there was likely pressure from husbands to withdraw from the association. Tribal identities came to the fore, reducing the willingness of some women to work in a collaborative manner, which had been so effective and enjoyable. Internal disputes arose, with instances of inappropriate behaviour by some women towards others, to the trainers and FPEC staff. A code of ethics (stating basic standards of behaviour) was introduced, but some refused to sign and left the association. The false rumours about the scale of funding available and how it was being used began to be stated openly as a matter of fact in social gatherings inciting locals. Consequently, a group of women 'hijacked' the March 2022 launch event to make their own speeches about the need for payments to women (irrespective of saleable products). The disputes and tension caused some of the able and effective FHWCA leaders of the project to leave Faynan for employment in Aqaba.

The UK funding stream ended in 2022. That was followed by a gradual period of withdrawal by FPEC while they continued to develop a plan for the sustainability and scalability of the endeavour. Although some of the local women leaders maintained their commitment to FHWCA through this period, pressures from the community and ultimately an unfeasible number of participants led them also to withdraw in March 2024.

3.4 Product Development

The market research for PEP established that the most financially rewarding products would be those that had the following qualities: unique to Faynan, utilised traditional craft skills, were high quality but low cost and were sufficiently small to fit into a tourist's luggage. A decision was made to use the geometric and zoomorphic designs found on Neolithic objects excavated from the site of Wadi Faynan 16 between 2008 and 2010 to create a series of products from pottery, textiles and copper sheet. Branding was devised based on the Faynan Heritage logo and a colour palette drawing on those from the Faynan landscape. The creation of a local brand was a difficult idea for some members of the FHWCA to grasp, as some felt that they could rely on reselling items that they purchased from other tourist sites in Jordan.

The copper products were identified as having the most potential because Faynan had been a centre of copper production from the Bronze Age to the Classical period, with many of its archaeological monuments relating to copper mining and smelting. The Neolithic sources consisted of stone arrowheads, geometric patterns on stone and bone objects, animal and human figurines. With the use of external consultants, a series of products were developed in a participatory approach with the FHWCA members including embroidered hoodies and T-shirts, woven mats, ceramic bowls, candle holders, copper pendants, brooches and bookmarks (Figures 3.2 and 3.3).

The women introduced their own products, including a version of the traditional Bedouin children's game known as tabb. Books on the archaeology and bird life of Faynan were donated by Mithen to the FHWCA for sale. Labels were produced to explain the Neolithic inspiration for each product and to specify the Faynan individual who had made it (Figure 3.4). Packaging – paper bags and cardboard boxes – was also designed by drawing on images of archaeological excavations and the Faynan landscape (Figure 3.5).

Product development faced three key challenges. The first was the low number of tourists between 2020 and 2023 and hence the quantity of sales. This arose from the Syrian Civil War being closely followed by the COVID-19 pandemic. These devastated tourism in Jordan, from which there was a very slow recovery up to 2023, which then collapsed owing to the Israeli–Gaza war. Without tourists, the

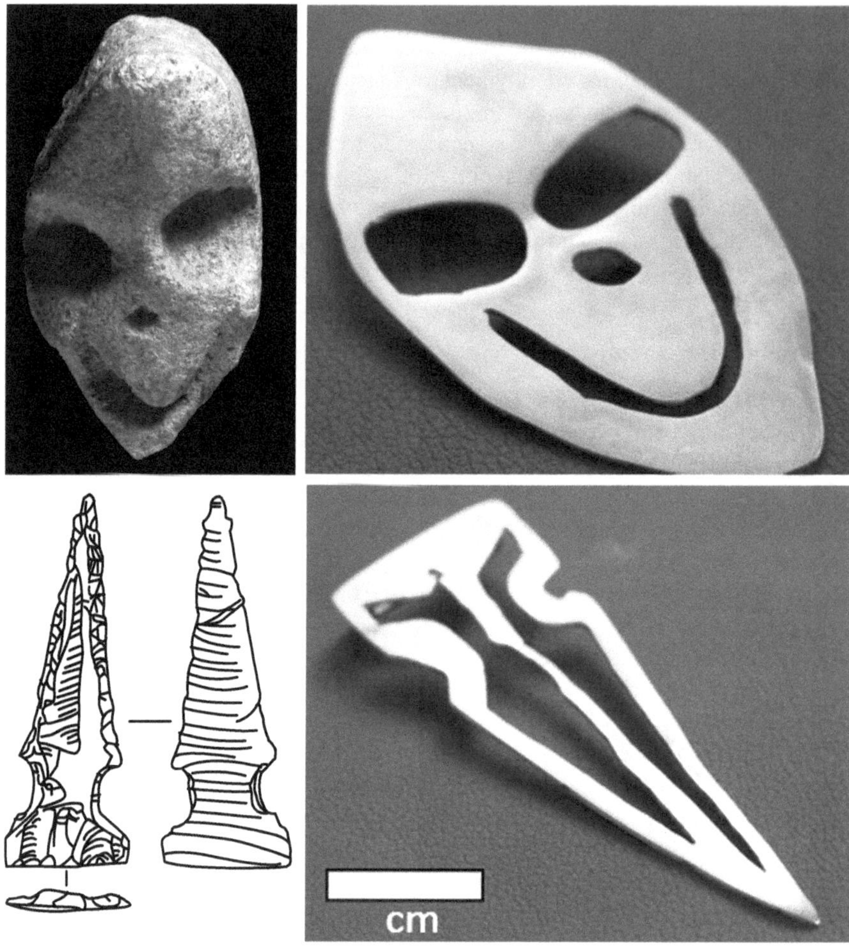

Figure 3.2 Neolithic stone face and El Khiam point from Wadi Faynan 16 translated into copper brooches. © Steven Mithen.

low sales made it difficult for the women to establish which products were the most appealing to visitors and struggled to maintain motivation.

Second, and associated with the first, was the culture shift required by the FHWCA members to make products that could be sold, rather than those they liked to make. The latter included various knitted garments such as table mats and bags. While suitable for their own homes and maybe a domestic market, these had little appeal to relatively wealthy international travellers.

Third, the quality of the products was often poor, especially for the prices being asked. While this partly reflected the time it takes to acquire new skills for pottery and metal working, it was difficult to instil the need for quality improvements

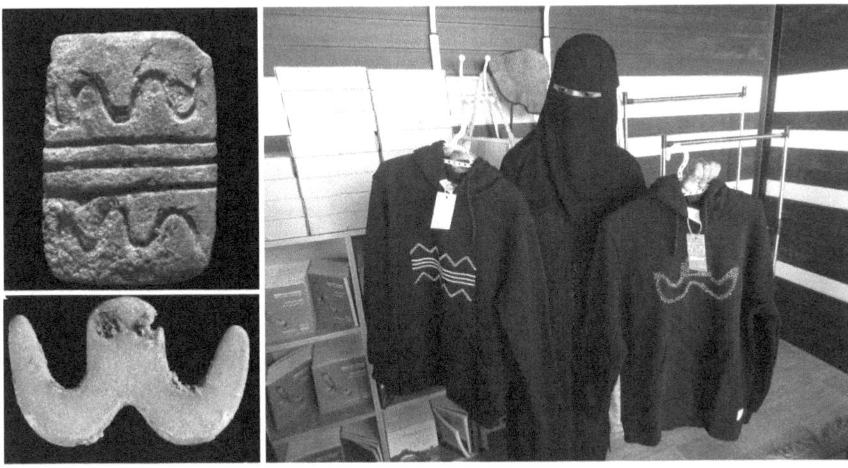

Figure 3.3 Neolithic incised stone plaque and shell pendant from Wadi Faynan 16 used as designs for hoodies. © Steven Mithen, with permission.

Figure 3.4 Tagged copper brooch and key rings. © Steven Mithen.

Figure 3.5 Gift box with a design based on excavation plan of Neolithic WF16, containing embroidered clothing. © Steven Mithen, with permission.

if sales were to be made and to persuade a more realistic pricing level. Overall, finding the balance between providing instruction and letting the local women have ownership of their products proved difficult.

3.5 Training in Crafts, Food Preparation and Business

The training programmes covered both craft and business skills and were delivered by FPEC staff and externally recruited trainers, with the provision of certificates to FHWCA members for completed courses. Regarding crafts, there was a focus on making pottery vessels. This was a new skill for the women, although one that had long been practised in Faynan as represented by scatters of ancient pottery in the fields and finds from excavations. Recruiting a trainer was more difficult than anticipated, with the first workshop delivered in September 2021 with 19 training days for 11 women. The women then worked by themselves, followed by a second workshop of 21 days in the spring of 2022.

A food service was also part of the plan for FHWCA activities. Although the kitchen provided was small, the aim was to cater for camps and tourists through specially developed local foods attractive and suitable for tourists. Such provision of food and drinks to visitors is a normal element of Jordanian and Bedouin hospitality and is often provided in gift shops. Moreover, Faynan lacks any cafes, restaurants, and suppliers of take-away food for visiting travellers. The women already had basic cooking skills, so the training focussed on menus that would be attractive to visitors and enhancing standards of hygiene in their preparation and service.

A substantial programme of business-related training was provided to the FHWCA leadership group and its members. The initial courses covered communication skills, compliance, people management, design thinking, product pricing, marketing and social media. The provision of training continued, responding to need. A three-day training course in accounting systems was delivered by an external provider in autumn of 2021 followed by one year of no-cost monitoring and support.

An external provider, LiveMean, was recruited to develop a website for FHWCA and an e-commerce platform for selling products online. The need for this became increasingly apparent as COVID-19 and international conflict led to a decline in visitors to Faynan. Once established, training was provided to selected women in the use and development of the sales platform. The required learning curve, however, was too steep, and their limited digital skills hindered and ultimately prevented the implementation of the online platform.

3.6 Infrastructure for Faynan Heritage Home

Following a period of assessing need and costs, it was decided that the HQ of FHWCA and Faynan Heritage Home would consist of three conjoined cabins designed to look like traditional Bedouin tents from the outside. This provided internal structured space for training, workshops, kitchen, dining, socialising and displaying items for sale. Following tenders, designer and supplier were commissioned to provide the cabins.

Securing land in a suitable position provided one of the most difficult and unanticipated challenges. Faynan gives the impression of empty space, but the politics and bureaucracy of land allocation were unexpectedly complex. The criteria for an optimal location were identified as being where: (1) tourists visit or have to pass through; (2) women will be safe and their equipment and products secure; (3) there is access to water and electricity; (4) women from all tribes could participate (i.e. external to any prescribed or recognised tribal area); and (5) women can readily access, ideally by walking a short distance.

The ideal location for Faynan Heritage Home would have been adjacent to the museum, creating a cultural heritage hub for Faynan. This was ruled out by the Department of Antiquities, apparently because of its policy against any commercial activity even though Faynan Heritage Home and the museum would remain entirely separate. Also, the museum's location in the Rashaydah village made it an unacceptable location for women from other tribes to regularly attend for work.

The views of the FHWCA members for alternative locations were surveyed, while discussions were undertaken with the Municipality of Qurayqira and Faynan and the Wadi Araba Development Company (WADC). Those discussions became protracted because the heads of these organisations frequently changed and always had conflicting views. Initial offers of land were made and then

Figure 3.6 The 'cabins' from exterior providing workshops, display area and sales point for Faynan Heritage Home, January 2022. © Steven Mithen.

withdrawn. A location on the edge of Qurayqira was finally agreed upon, one that sufficiently met the location criteria. After another lengthy bureaucratic process to secure Ministry approval, and considerable preparatory work (levelling ground, securing electricity and water supply), the cabins were installed in the spring of 2021 (Figure 3.6). That was a milestone for the project, providing the workshops, product display and sales point for Faynan Heritage Home.

In November 2021, the MoTA informed FHWCA that the location of the caravans would have to be moved because the existing site was prone to flooding. Neither FPEC nor FHWCA could find any reason for that, the location being no different to all others in the vicinity of Qurayqira. The MoTA agreed to compromise by allowing a retaining wall to be constructed around the cabins, but no funds were available to do so. As such, a new location had to be identified and the cabins moved, inflicting additional cost and causing a further delay to the business development.

The final location was on the outskirts of Qurayqira, and within the walled compound of an existing community association, the Al Sharif Nasser Association (ANA). This had some advantages – the locked gates aided security, water and electricity provision were improved, and an empty building could be used for the pottery workshop (Figure 3.7). It was, however, further off the 'tourist trail' with limited visibility behind the compound wall and further from the homes of the FHWCA members. That resulted in FHWCA members requiring additional transportation fees, which were paid to male members of their families who drove them to the cabins.

Further, the relationship of the FHWCA and the Al-Sharif Nasser Association (ANA) started to become tense, because the use of the premise was interpreted by some members as a form of ownership. Some members of the ANA believed that

Figure 3.7 Yusra Al-Jabari, project manager, viewing newly made Faynan heritage ceramics in the pottery workshop, January 2022. © Steven Mithen, with permissions.

Figure 3.8 Inside the cabins, March 2022, showing display area and working space. © Steven Mithen.

the FHWCA should be merged with the ANA, because they were now working from the same space. The previously mentioned rumours concerning the level of project funding were also at play, and suggestions for large payments were made in return for the use of the location and premises. ANA members sometimes attempted to put constraints on FHWCA access, water use or land use. Several talks between the FHWCA and ANA leadership groups in the presence of regional or national tribal leaders repaired the relationship and secured the use of ANA compound for the FHWCA location without further trouble.

The cabins and adjoining rooms were equipped with furniture, kitchen items, workshop equipment, shelving for display of products and facilities for making sales. (Figure 3.8) As the work progressed, the equipment was enhanced with a new pottery kiln and copper working tools provided in the summer of 2023.

3.7 Marketing, Publicity and Sales Outlets

FPEC staff worked with the FHWCA leadership group to develop a marketing strategy and identify sales outlets. FPEC used its network to put the women into contact with the Wadi Araba Development Company, the Directorate of Culture in Aqaba and the Ministry of Culture in Amman so that the FHWCA would receive information about opportunities. FPEC facilitated visits to Faynan by the Director General of the Directorate of Tourism and developed links on behalf of the FHWCA with the Directorate of Education, following interest that its members had shown in the FHWCA products during a visit to Faynan (especially the books and the tabb game).

Considering the less-than-ideal location of the cabins, other potential sales outlets in Faynan were explored, notably the trekking camp at Wadi Ghuwayr. The headman, Abu Majid, was supportive, but the women needed to attend the camp at the specific times that the trekking groups were present. The FHWCA women declined this opportunity due to family restrictions on their movements in Faynan. FPEC designed signage for placement at the Ecolodge, trekking camps, the Ecolodge reception centre and on approach roads to direct visitors to the FHH cabins.

Restrictions and/or reservations held by the FHWCA women about movement outside of Faynan required the FPEC staff to pursue sales outlets in Amman, Aqaba and Wadi Musa (Petra). Some outlets were pleased to accept the FHWCA products, and the contacts were handed over to the FHWCA to pursue stressing the need for them to be proactive with realistic pricing policies. A connection was made with the Al-Disa Women Association in Ghour Al Safi, enabling FHWCA to display their products at the Al-Disa bazaar in April 2023.

Marketing, however, proved to be one of the most challenging components of the project. The FHWCA women lacked the skills for seeking new outlets and for pursuing established cooperation possibilities. Within a culture that has severe limitations on women's movement, they found themselves completely unable to promote their products or obtain direct feedback from consumers. This was a major issue due to the relatively isolated location of Faynan, which makes any movement costly even within the region. None of the participating women owned any vehicle, and there is no public transportation within the area. The outlets created by FPEC provided support but that were insufficient in number and not adequately maintained by the FHWCA women to achieve significant sales. The problems arising from limited desire and ability for women to travel out of Faynan were never solved and became a major issue preventing the sustainability of the Faynan Heritage Home.

3.8 Support and Mentoring from FPEC

The complex processes of creating a legal entity, securing land and establishing a business were undertaken by FPEC on behalf of the FHWCA. Throughout this process, FPEC provided close support and careful monitoring of the FHWCA. Their Amman office is 200 km from Faynan, and travel had been legally restricted during the COVID-19 pandemic years of 2020 and 2021. Whenever possible, FPEC members, notably Nebras Maslamani and later Yusra Al-Jabari, visited Faynan on a weekly basis and sometimes biweekly. These were supplemented by daily contact via a WhatsApp group, phone and online meetings. FPEC also helped the FHWCA prepare its own applications to funding bodies. It achieved one success with a small grant from the Ministry of Culture for a sewing workshop.

As the training was completed and the experience developed, FPEC sought to transfer responsibilities to the FHWCA itself. Nevertheless, even in 2023 it had to provide support by reviewing sales lists and gathering feedback from points of sale, monitoring the paperwork relating to each member, maintaining inventories of assets including products, raw materials, equipment and preparing invoices and financial records. FPEC continued to provide advice about improving the quality, reducing the cost of the products and developing relationships with points of sale. Overall, the shift of responsibilities to establish the FHWCA as an independent, self-managing entity was slower and more challenging than had been experienced in similar projects elsewhere in Jordan.

3.9 Reflections

At the time of writing (October 2024), the FHWCA and its business arm Faynan Heritage Home cannot be described as a success because the activity has effectively ceased. Three key reasons can be identified for this, which may provide lessons for devising similar projects in Jordan and elsewhere.

Timeframe and Funding: The PEP project was funded by two grants from the UK, an AHRC award with a spending time frame of 12 months and the Newton Prize with that of three years. Supplementary funds for 6 months were secured in 2023 that allowed maintenance of the cabins and enhanced workshop equipment, but the funding period was too short and the amount of funding too low for the ambition of the project. The number and complexity of the bureaucratic processes involving several ministries, directorates, and municipal bodies were significantly underestimated, as was the extent of culture change required within the Faynan community. The key lesson is that the scale and relatively short-term funding provided to academics in the UK are insufficient for projects, which require a long-term development process. This reflects the UK academic environment of 2018–2023 that prioritised short-term transitory returns over long-term achievement. In the case of Faynan, funding support spread over ten years would have been required to ensure a sustainable business was developed.

COVID-19 and Conflict: Project development was severely impacted by the COVID-19 pandemic that inhibited activities during its early stages and left a long legacy, notably a drastic decline in tourism. That was exacerbated by the conflict in the wider region, notably the ongoing conflicts in Syria (since 2011) and Israel–Gaza (as from October 2023). Without adequate tourism, the FHWCA was unable to sell sufficient products to generate income and learn which products were the most rewarding to manufacture; without financial returns, women became unmotivated to enhance their products and undertake the required work. It would be wrong to take the lesson that such projects should not be undertaken in areas of long-term conflict and tension such as the Middle East. Instead, they require embedded contingency plans for 'events', a larger scale of funding and longer timeframe for expenditure than projects undertaken in more stable environments.

Society and Culture: With its four Bedouin tribes, Faynan was known to be a challenging area to work within, as had been experienced during the museum development. The extent of this, however, was underestimated with a surprising amount of social tension between individuals and groups of women, partly arising from that between the tribes. That impeded the training and ability of FPEC to provide support for its advice to be acted on. Conservative social attitudes constrained the women from taking the necessary initiatives to market and sell their products and – we suspect – fully engage in the association. A long succession of academic and development-related projects in Faynan appeared to have created expectations of ongoing funding, with an inexhaustible supply of grants from the UK. Those expectations could not be replaced with an understanding of the need for a commercially viable enterprise, reliant on the income that it could generate. The key lesson is not only the need for more time, but also the need to invest resources into communities who not only request support for business development – as did the Faynan community in October 2018 – but also are socially and culturally ready to work constructively to achieve positive outcomes. Unfortunately, that was not the case for the Faynan community.

3.10 Conclusions

The Faynan Heritage Project has sought to use cultural heritage to support eco-tourism into a disadvantaged region of Jordan for the benefit of the local community. The project developed piecemeal, with two main initiatives: developing a museum (this volume, Chapter 2) and a handicraft business owned, managed by local Bedouin women, as described in this chapter. The idea for a handicraft business and request for support came directly from the Bedouin women and Faynan community leaders themselves. It was undertaken via a partnership between a UK University that secured and managed the funding, a Jordanian University with its PI, Fatima Al-Namari, having experience in similar projects, and a not-for-profit organisation with extensive experience of working with disadvantaged communities in Jordan and success with developing similar projects elsewhere in the country. Extensive effort and resources were placed in working in full collaboration with the Bedouin community throughout the project, primarily via the work of Nebras Maslamani and Yusra Al-Jabari.

Despite initial success, the ultimate failure of the business to develop can be attributed to three main factors: available funding was insufficient in quantity and was available over too short a timeframe for the required tasks, those involving Jordanian government ministries being far more complex than anticipated; the project coincided with COVID-19 and international conflicts that decimated tourism into Faynan; despite its initial request, the social structure and culture of the Faynan community were unable to undertake the work and adopt the mind frame required for business development within the timeframe available.

Nevertheless, times change. Faynan Heritage has left foundations in place for heritage-based ecotourism development, not only in the infrastructure of the museum and walking trail, but also from the training provided to the women and activities delivered to the schools and children. We hope these may deliver rewards during more advantageous times.

Acknowledgments

The authors are grateful to the UK AHRC and Newton Fund for the provision of funding that supported this project. Staff at the University of Reading and FPEC offices helped administer the funding awards. We thank all the Jordanian bodies who provided their approval and/or support for the project, notably the MoTA, the Ministry of Culture, the Municipality of Qurayqira and Faynan and the Wadi Araba Development Company. Finally, we thank the local community in Faynan, especially all the women who participated in the project and more particularly those who showed leadership and commitment. We enjoyed working with you, learned from you and hope to do so again in less troubled times.

Notes

1 Lina Annab, quoted by Oxford Business Group, 15 February 2017, https://oxfordbusinessgroup.com/news/positive-momentum-established-jordan%E2%80%99s-tourism-industry (accessed September 2017), and quoted in *The Jordan Times*, 24 July 2017.

2 *Arts and Humanities Research Council* for 'From Archaeology to Business in Faynan: Empowering and alleviating the poverty of Bedouin women in southern Jordan' AH/T00777X/1.

3 BEIS R&I ODA programmes, NP2020PB\100028.

4 Future Pioneers for Empowering Communities (FPEC) is a Jordan-based non-profit organisation specialised in community development with a focus on women, youth and marginalised groups. Full details can be found on its webpage: https://future-pioneers.org/

4 Community-Based Conservation and Promotion of the Neolithic Site of Beidha, Jordan

Bill Finlayson

4.1 Introduction

Neolithic Beidha, a short walk from Little Petra, was largely excavated in the 1950s and 1960s by Diana Kirkbride, with one single later season in 1983, initially under the sponsorship of the British School of Archaeology in Jerusalem (BSAJ) and by the then British Institute in Amman for Archaeology and History (BIAAH) for the final season (Figure 4.1). Kirkbride had worked at Jericho with Kathleen Kenyon, where the Neolithic was found in deep trenches through the tell, limiting insights into how such an early settled community had lived. At Beidha, Kirkbride was able to expose a considerable part of the settlement, providing a previously unseen picture of Neolithic lives. Her open-area excavation subsequently became the norm for Neolithic excavations in the region. The remains of some 65 well-preserved Neolithic structures in three main phases were uncovered, with the excavation attracting considerable attention as it appeared to show the evolution from circular to rectilinear architecture, an important part of the Neolithic story about settling down and the start of village lifestyles. Kirkbride's excavation became well known for its pioneering nature, its contribution to an important part of human history, and for the remarkably good preservation of the settlement (see Byrd 2005 for the main excavation report and references therein).

The Beidha Project started in 2001, not long after I became director of the Council for British Research in the Levant (CBRL), the successor to both the BSAJ and the BIAAH (Dennis et al., 2002; Finlayson et al., 2003). There were both push and pull factors behind starting the project. Push factors were mainly related to the deteriorating architecture that had remained exposed at the site since Kirkbride finished excavating in 1983. As was typical at that time, Beidha had been left as it was on the day that excavation ended, providing an opportunity for people to visit the site but with neither conservation to protect the site nor visitor interpretation (Kirkbride, 1984). The expected number of visitors was small, because anyone managing to find Beidha could be assumed to have some knowledge of the Neolithic, with no need to cater for popular tourism. What visitors found, however, was a deteriorating archaeological site.

The remains of the Neolithic village were surrounded by spoil heaps and the deep empty trenches that had defined the limits of the settlement. An ugly

DOI: 10.4324/9781003491071-5

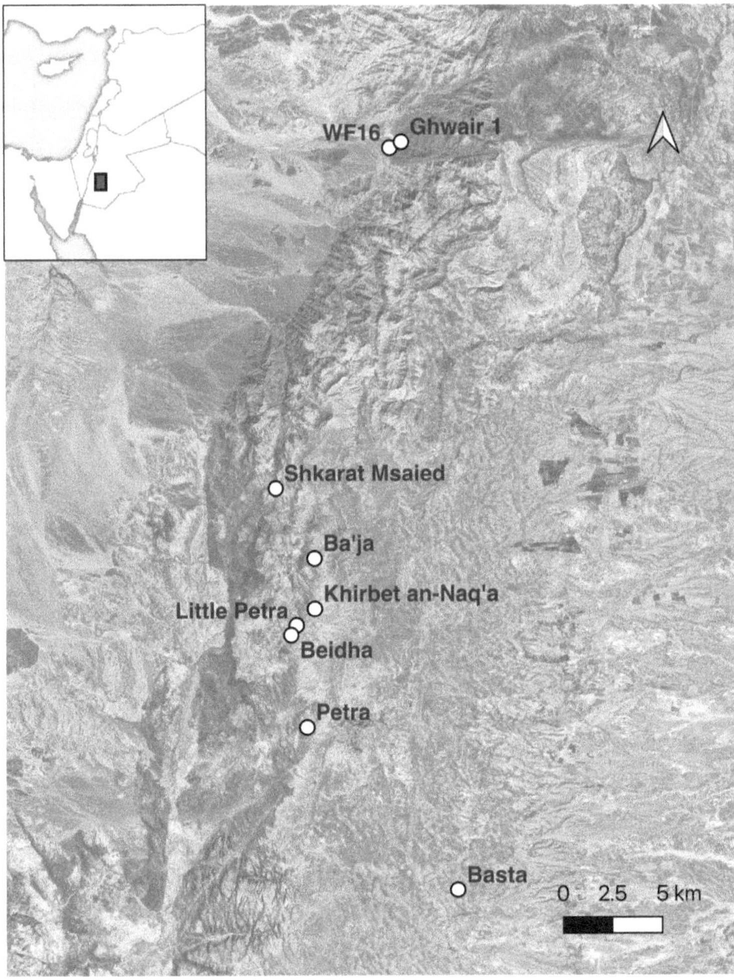

Figure 4.1 Map showing the location of sites mentioned in the text. © Bill Finlayson.

chain-link fence had been put up around the site and dominated the skyline. Apart from marking its location in the landscape, the fence served no practical function because its gate was never locked, allowing goats to easily enter. These issues affected the overall visual impression of the site. The once well-preserved Neolithic remains were also deteriorating. Winters in Beidha can be harsh, and seasonal damage had been taking its toll; wall collapse at Neolithic sites can be rapid after water penetrates the mud mortar used between the stones.

Kirkbride left several pillars of unexcavated material, to both show the depth of deposits that had been excavated and serve as a potential resource for later archaeologists with new analytical techniques. Despite such good intentions, these tall narrow pillars had eroded more rapidly than the exposed remains below and started to collapse, knocking over the *in situ* walls. Soil accumulating within the trenches from the collapsed pillars and as windblown sand was allowing scrubby vegetation to grow, both masking remains and further damaging the architecture by its invasive roots. The few tour guides who brought visitors to the site would often walk over the Neolithic architecture, slowly but steadily damaging the tops of the walls and surviving floors. In short, the site was in a mess and the speed of dilapidation was accelerating. As Director of the CBRL, I had inherited responsibility for addressing these issues, made more immediate by my own interests in the Neolithic.

My interest in the Neolithic meant I desired not only the protection of the site, but also its public promotion. That was part of a wider concern that the world-class Neolithic remains in Jordan were a neglected part of its cultural history for both Jordanians and international visitors (Finlayson et al., 2015). Aysar Akrawi, then Director of the Petra National Trust (PNT), was also concerned about the state of the site and had an additional concern for the Amarine Bedouin tribe, many of whom lived in the nearby modern village of Beidha and who had been increasingly excluded from the economic benefits of the Petra World Heritage Site. The late Ali Jabri, an artist with a deep affection for Jordanian heritage and archaeology, was another influential advocate for the site and became a regular volunteer in our early efforts. Dr Mohammad Najjar, then Director of Excavations at the Department of Antiquities, whom I had grown to know through our shared Neolithic interests and joint work in Wadi Faynan, became another enthusiastic supporter.

The context of Neolithic Beidha had also changed because throughout the late 1980s and 1990s cultural heritage tourism had become increasingly important to the Jordanian economy. In 1985, Petra had been inscribed on the World Heritage list, and both the road infrastructure and tourist accommodation in nearby Wadi Musa had developed – processes that have continued ever since. International recognition of the Jordanian Neolithic had also grown with increasing numbers of excavations, including many in the wider Petra areas such as Ba'ja, Basta and Shkarat Msaiad; Neolithic objects from Jordan were now on display in major museums around the world, notably from 1997 the 'Ain Ghazal statue in the Louvre, with additional statues now on display in the British Museum. Efforts were being made in the Jordanian tourism industry to find ways of persuading tourists to spend longer in the country – too many visits were very short to only see the highlights, with very few repeat visits.

4.2 The Beidha Project

With support from Aysar Akrawi, Ali Jabri and Dr Mohammad Najjar, the Beidha Project was established in 2001. One approach to preserving and promoting Neolithic Beidha would have been to apply for major funding to erect a cover across the whole site. That was resisted, however, because too much cultural heritage work has become driven by funding rather than by the requirements of the heritage, the local community and visitors. We noted situations where, once major funds had been secured and spent, interest in the now protected (but largely hidden) heritage sites had rapidly evaporated. For instance, the enormous infrastructure development project at Umm ar-Rasas, inscribed on the World Heritage list in 2004, took place in parallel with our work at Beidha, but neither visitors nor the local community have gained any benefit.

Our alternative approach at Beidha was to work at a small scale and engage people from the local, regional and international communities in the prehistory of Jordan, seeking to attain a longer-term and more self-sustaining impact. We did not wish to build another monument, such as another visitor centre or even a shelter, but to develop human interest and knowledge; moreover, a cover across the site was undesirable because it would be visually intrusive in the landscape and bring its own problems of drainage and maintenance.

Our initial efforts were low key and often conducted on weekends. Samantha Dennis, then working as a research scholar at the CBRL, joined us and then took charge of an experimental reconstruction programme that she completed as part of her PhD at Edinburgh University (2008). Other volunteers from the CBRL, including Charlotte Schriwer and Sam Smith, and team members from our Neolithic projects in Faynan and at Dhra' all joined the volunteer task force. We employed many of the local Amarine, especially Mohammed Ibrahim, Faisal Abdul Aziz and Musa Abdul Aziz who became Samantha's regular team, while others worked on a voluntary basis. We employed an additional site guard, Ms Asmahan Ali, who enthusiastically conducted basic site care in her spare time, before leaving us to get married. Ziad al Saad let us use his house in Beidha, before developing the Amarine camp which became our base for a while.

Initial work was quite simple: moving the chain-link fence from the skyline and repairing it to keep goats away from the site, removing the ring of spoil heaps that hid the site and using the material to fill in the empty deep trenches or to batter the collapsing vertical sections to stop their erosion and bringing down the collapsing pillars in the site (Figure 4.2). An experimental reconstruction programme was started – intended from its inception to serve both as an archaeological experiment and to help interpret the archaeological remains for visitors (Dennis, 2003a; 2003b; 2008; Finlayson et al., 2006). Team members gave lectures at both academic and public events and a presentation to the Jordanian Tour Guides Association. The team also contributed to PNT primary school events in the villages around Petra. In 2003, a path was created around and through the site, designed to keep guides and their groups off the archaeological remains. Temporary interpretative signs in Arabic and English were erected. The reconstructed Neolithic village became the set for several TV documentaries on the Neolithic, such as the 2005 Lion TV production of Guns, Germs and Steel for National Geographic.

Although Beidha will never attract the large crowds who visit Petra, increasing numbers began to make the short walk from Little Petra. Some were led by guides, while others were pleased to escape their guides (who saw Beidha as an unnecessary stop) or were independent travellers. Many of these visitors were well informed and interested, happy to chat about the difficulties of conserving a site where the stonework was held together by a mud mortar prone to washing away.

Figure 4.2 Initial site works showing the excavations. The fence on the skyline has been removed, and empty trenches have been filled (one is still visible as a darker square of newly moved soil). The collapsing pillars can still be seen in the site, and the volunteers are moving the last of the spoil heaps that previously surrounded the exposed remains. © The Beidha Project.

4.3 A Neolithic Trail

By 2007, the project at Beidha had attracted considerable attention. To expand the concept of community-led heritage management, a Neolithic Heritage Trail was proposed to the Jordanian Department of Antiquaries (DoA) and the Ministry of Tourism, in partnership with archaeologists working at nearby Neolithic sites (Baja, Basta, Ghuwayr 1, Shkarat Msaied and Wadi Faynan 16) – notably Hans Georg Gebel, excavating at Ba'ja and Basta, but also Charlott Hoffman Jensen and Ingolf Thuesen of Copenhagen University, Alan Simmons from the University of Nevada at Las Vegas and Talal Hamad al-Amarin, a local Bedouin educated as an archaeologist. This followed the principles of the Beidha Project and was intended to provide an opportunity for visitors to see the Neolithic sites running from Beidha to Faynan, encouraging stewardship of the sites by the local Bedouin communities. With the newly constructed wilderness lodge by the Royal Society for the Conservation of Nature (RSCN) in Faynan (later named as the Ecolodge, in this volume, Chapter 2), extensive accommodation in Wadi Musa, walking routes and an improved road in the Wadi Nimala, the logistics of providing a Neolithic Trail appeared reasonably easy. Hans was investigating the possibility of establishing an al-Ammarin Heritage Center, together with the local head of the al-Ammarin Cooperative at Beidha. The idea was to use Khirbet an-Naq'a, the nearby abandoned old village of nine houses, about 1 km from Beidha on the road between Wadi Musa and Wadi Araba to establish a small museum, a handcraft centre and a Neolithic research station. A study project by a member of the Ba'ja Neolithic Project, Maryam Bshesh, as part of her master's programme at Cottbus Technical University, Germany, investigated a marketing strategy for the Neolithic Heritage Trail, following sustainable principles. Not all these ideas were realised, but the idea of a Neolithic trail was planted and at an informal level was followed.

Fortunately, the Neolithic trail proposal coincided with other tourism initiatives to develop trails and hiking, some within a USAID-funded heritage tourism development programme. Following a meeting with its director, Ibrahim Osta, an MoU was signed in 2010 between USAID/Jordan Tourism Development Project II (JTD II), the Petra Development and Tourism Regional Authority (PDTRA) represented by the Petra Archaeological Park (PAP), the Department of Antiquities (DoA) and CBRL, to further develop and conserve the Neolithic site at Beidha. This provided the first external funding and allowed Dr Pascal Flohr to join the project. She repaired the experimental reconstructions, made them safe for visitors (Figure 4.3) and supervised the establishment of a more permanent path and ceramic signs of the type now being used in Petra (Flohr et al., 2015).

A review of previous backfilling of trenches and conservation led to additional backfilling using imported sand. A walking route from Little Petra was established, with low-cost concrete signs along its length marking out a timeline going back into the past showing milestones of human achievement and local historical events. The initial plan had been for these signs to be incorporated on large natural rocks which would have been more robust, but the cost was prohibitive.

Partial backfilling, fencing and the installation of another ceramic sign were undertaken at Shkarat Msaied on the Wadi Nimala Road. The final fieldwork at Wadi Faynan 16 in 2010 involved the construction of a replica Neolithic structure within an easy walk from the RSCN wilderness lodge (Flohr et al., 2015). More work has since been undertaken in Faynan, including the opening of a DoA museum, as reported elsewhere in this volume (Chapter 2).

Importantly for the project, the PDTRA agreed to undertake the longer-term maintenance of Beidha, both the archaeological remains and the experimental structures, a responsibility undertaken assiduously until COVID-19. The only part of the programme that started to fail was the low concrete signs on the walk in from Little Petra. These became damaged by vehicles and, although initially propped back up, were eventually abandoned.

The trail project had proposed the use of local Bedouin as guides. Many of them had worked on excavations, and this seemed a good way to help encourage local interest and concern for the sites. The difficulty lay in their access to higher education, a requirement for becoming official guides and an issue recognised by the PAP by at least 2011. There was little direct success, but some of the local Beidha

Figure 4.3 The reconstructions and paths tidied under the USAID project, before the installation of signs. © The Beidha Project.

Bedouin began to talk about the archaeology when they were taking visitors hiking. Similarly, some of the RSCN-trained Bedouin guides based in Faynan began to include the archaeological sites on their hikes.

4.4 Seeking Sustainability at Beidha and Beyond

In 2014, I conducted a season with Cheryl Makarewicz to refresh the experimental buildings and pathway while excavating a larger structure in the centre of the site. The long-term maintenance plan and its ongoing success were discussed with the Petra Archaeological Park. Strikingly, it was clear that the number of visitors to Beidha had increased enormously with many groups including it on their schedule and thousands of visitors passing by requiring students working on the project to be assigned as guides.

Figure 4.4 Oroub El-Abed talking to a group of Amarine school students visiting the site. © The Beidha Project.

In 2015 with British Academy funding, Oroub El-Abed, Paul Burtenshaw, Carol Palmer and I conducted the *Deep Past as a Social Asset* Project to explore the relationship between the modern communities of Beidha and Basta and the archaeological sites they lived around (El-Abed et al., 2018; Burtenshaw et al., 2019). Using both qualitative and quantitative surveys, we examined how people valued these archaeological sites, for their cultural value, or as potential economic resources. In addition to the direct research, public events, group meetings with local school students (Figure 4.4) and a workshop in Amman were held. New bilingual (Arabic and English) information leaflets were produced for the Neolithic sites and the Neolithic trail. This project led to new relationships with the communities and increased understanding of the local context.

The community living around the larger Neolithic settlement at Basta, located outside the Petra Park boundaries, was generally not keen on the development of international tourism and the disruption to their community it might entail. They were more interested in the abandoned old stone village which figured in family stories and the possibility of Jordanian tourism with local visitors coming to see their still-functioning traditional gardens and farming and buying the dairy products made in the village. Many were enthusiastic about backfilling the Neolithic remains, where deep trenches and fragile walls were seen as a hazard for children and prone to filling up with rubbish.

In contrast, people at the more recently established settlement at Beidha, within the Petra Park, appeared to have a greater interest in antiquity, especially the Nabatean past about which they have a sense of ownership. Many of them already received some economic benefit from the heritage industry and could see the potential benefits arising from more heritage tourism. However, while ideas have been put forward, there have been few local initiatives to follow these up and a lack of engagement by the DoA and local government bodies. Although archaeology is now in the school curriculum, teachers have little time available to make any field trips or to visit museums. As Burtenshaw had found during his PhD research in Wadi Faynan, knowledge of archaeology rarely spread beyond those working directly on the sites.

Local interest is not only important for site stewardship, but also to develop sustainable tourism throughout Jordan. The rise and fall of Middle Eastern security problems, while rarely impacting Jordan directly, has frequently had a major effect on its tourist industry. We started thinking about presenting Beidha in 2001 and Samantha Dennis began market research on visitor interests. Unfortunately, at that point the tension with Iraq that led to the 2003 US invasion meant that tourism to Petra was already collapsing. That cycle continues, now (2024) with the Gaza crisis.

In the AHRC Newton Khalidi-funded 'Rewriting the Prehistory of Jordan' project, the team has attempted to increase knowledge of prehistoric archaeology in Jordan, professionally with the DoA, academically with graduates from Yarmouk University, and to a wider public by producing a book targeted at secondary school children. Written by three Jordanian members of the team, Professors Zeidan Kafafafi, Maysoon al-Nahar and Dr Sahar Khassawneh, this heavily illustrated book in Arabic on Jordan's prehistory is to be distributed by the Ministry of Education in 2024/2025. At present, the popular conception of prehistory is often that it is only of interest to foreigners. We are trying to challenge that, to show how important prehistory, especially the Neolithic, is to the human past and to modern Jordanians; we wish to emphasise how the Neolithic remains in Jordan are an important part of global history.

4.5 Reflections

The Beidha Project has been an interesting journey so far. There have been several striking successes, notably the conservation of the site, the use of replicas for experimentation and interpretation and a wider exposure of people to the Neolithic. There have also been some failures, not least that the project has required constant injections of activity. It has been harder to maintain activity since I relocated to the UK in 2012 and ceased working for the CBRL in 2018.

During a 2022 visit to Beidha, it was clear that the maintenance schedule had collapsed, possibly because of COVID-19. Routine, low-level maintenance had been critical to keep the experimental buildings intact and to make sure the low-intervention conservation measures were kept in a good state or repair. The intention was to ensure that with a small annual maintenance effort no major new work would be required. The signs, which for a long time had been untouched, had been vandalised. It seems clear that this was an inevitable consequence of the authorities no longer showing interest in the site. There had been one further intervention on the site, when a UNESCO-sponsored project had usefully installed some drainage features to keep water away from the site. Unfortunately, that project had also started invasive 'conservation' work on some buildings. The work appeared to have been stopped suddenly probably due to its controversial nature. When I returned with a group of DoA trainees from the 'Rewriting the Prehistory of Jordan' project later in 2022, we reinstated the paths and structures.

The UNESCO-sponsored work had been part of a wider development for Petra, creating a route for shuttle trucks to enable tourists to enter Petra on the far side of the site, away from the main entrance, thereby helping reduce pressure on that entrance. This route had been taken by walking groups in the past, many of whom had stopped to visit Beidha. Tourists could now walk from the car and coach park at Little Petra to wait for a shuttle to collect them from an area close to the Neolithic site. Unfortunately, the remaining signs to the Neolithic site had now been lost and there was nothing to guide visitors from the shuttle truck waiting area to Beidha. Some of those who came to see the site while we were working there were still faced with the old difficulty of guides on a strict schedule, not wanting their groups to stop and look at Neolithic remains.

Tourism at Beidha falls between two models of tourism. The main attraction of Petra is a 'hot spot' where tourists and tourism investment are concentrated (Robinson, 2001; Sandrine, 2005). When the Beidha Project was being developed the problems with this model appeared significant, in that it allowed tourist visits to be brief and for the economic benefits to be concentrated within the 5-star hotel industry (Daher, 2006). Equally problematic was that such concentration around hot spots creates a form of enclave tourism where actual cultural experiences could be highly sanitised and artificial. The practice of some guides to rush their groups from selected experience to selected experience and then to an approved restaurant with little flexibility to take in new additional experiences is best understood in this context. The Beidha Project was a conscious attempt to help slow down tourists, encouraging them to spend longer in the country, in a way that coincidentally fell in line with the USAID Siyaha promotion of trails. The encouragement of local, Jordanian tourism similarly was intended to help develop a form of tourism less open to the boom-and-bust economy of international Middle Eastern tourism.

As this chapter was being completed (March 2024), at the launch of the report 'Preserving Legacies: A Future for Our Past' (on the impact of climate threats at Petra), a representative from the PDTRA stated that because of the ongoing conflict in Gaza and associated troubles, visitor numbers at Petra were down to about 1,000 a day which would not bring in sufficient revenue to maintain their Petra management programme. Sadly, while Jordan remains a very safe country, such regional disruption is a normal part of the tourism cycle. Although local tourism, walking trails and smaller sites will never bring in the revenue of a Petra, they are a more sustainable form of tourism that permits a more genuine cultural exchange.

It seems that both local and international champions are still required to promote individual sites. The Beidha Project members continue to maintain good relationships with the DoA and PDTRA. At the community level, despite attempts to build local interest and capacity, it has been hard to sustain this, as the archaeological job market has – outside the major institutions – been too informal with employment often poorly rewarded and insecure. Sela/Alraqeem, a Jordanian grassroots organisation dedicated to building capacity for local-level engagement, is attempting to change this situation (Øystein, LaBianca and Ronza 2023; this volume Chapter 5). Sela has already been engaged in erecting signs in Wadi Faynan and could be part of the future of the Beidha Project and the Neolithic Heritage Trail.

Acknowledgements

There are too many people to acknowledge all the individuals whose help has facilitated this long-running project. Colleagues who have made significant contributions to the site presentation and tourism project include Mohammad Najjar, Oroub El-Abed, Samantha Dennis, Pascal Flohr and Paul Burtenshaw. As a community project, I have to thank all the many Amarine who assisted the project in many ways over the years – and I hope we can continue this partnership. Of course, all the Directors General and the staff at the Department of Antiquities. From other projects, Hans-Georg Gebel and Moritz Kinzel have frequently engaged in supportive discussions.

References

Burtenshaw, P., Finlayson, B., El-Abed, O. and Palmer, C. (2019). Using the past for local community futures in Jordan, *Conservation and Management of Archaeological Sites*, 21(2), pp. 69–91.

Byrd, B.F. (2005). *Early village life at Beidha, Jordan: Neolithic spatial organization and vernacular architecture British Academy Monographs in Archaeology, CBRL*. Oxford: Oxford University Press.

Daher, R.F. (2006). 'Chapter 1. Reconceptualizing tourism in the Middle East: Place, heritage, mobility and competitiveness', in: R. F. Daher (ed.) *Tourism in the Middle East: Continuity, Change and Transformation*. Bristol, Blue Ridge Summit: Channel View Publications, pp. 1–69.

Dennis, S. (2003a). The experimental reconstruction of a Pre-Pottery Neolithic B structure at Beidha—a visual introduction, *Levant*, 35, pp. 39–48.

Dennis, S. (2003b). PhD dissertation project: The use of experimental archaeology to explain and present Pre-Pottery Neolithic architecture at Beidha in southern Jordan, *Neo-Lithics* 2(3), pp. 37–38.

Dennis, S. (2008). *The Use of Experimental Archaeology to Examine and Interpret Pre-Pottery Neolithic Architecture: A Case Study of Beidha in southern Jordan*. Ph.D. University of Edinburgh.

Dennis, S., Finlayson, B. and Najjar, M. (2002). Conservation and presentation of Neolithic Beidha, southern Jordan, *Antiquity*, 76, pp. 933–934.

El-Abed, A., Finlayson, B., Burtenshaw, P. and Palmer, C. (2018). *The Deep Past as a Social Asset in the Levant (DEEPSAL): A study on the livelihoods of the local communities living close to the Neolithic sites of Basta and Beidha in southern Jordan, Council for British Research in the Levant, funded by the British Academy.*

Finlayson, B., Burtenshaw, P. and Palmer, C. (2015). The Deep Past as a Social Asset in the Levant, *British Academy Review*, 25, pp. 18–23.

Finlayson, B., Dennis, S. and Najjar, M. (2006). *Beidha Experimental Archaeology CBRL 2006, Bulletin of the Council for British Research in the Levant, CBRL*, pp. 22–23.

Finlayson, B., Najjar, M. and Dennis, S. (2003). Bayda: Conservation and presentation project, April-November 2003, *Annual of the Department of Antiquities of Jordan*, 47, pp. 489–490.

Flohr, P., Finlayson, B., Najjar, M. and Mithen, S. (2015). Building WF16: Construction of a Pre-Pottery Neolithic A (PPNA) pise structure in southern Jordan, *Levant*, 47(2), pp. 143–163.

Kirkbride, D. (1984), Beidha 1983: An interim report, *Annual of the Department of Antiquities of Jordan (ADAJ)*, 28, pp. 9–12.

Øystein S., LaBianca and Ronza E. (2023). Site custody activism in Jordan: introducing *Sela* and *Alraqeem*, *Levant*, 54(3), pp. 444–450.

Robinson, M. (2001). 'Tourism encounters: Inter- and intra-cultural conflicts and the world's largest industry', in: N. Al-Sayyad (ed.) *Consuming Tradition, Manufacturing Heritage: Global Norms and Urban Forms in the Age of Tourism*. New York: Routledge, pp.34–67.

Sandrine, G. (2005). Shaping the city for the other: Tourism development, heritage management and the urban fabric in Luxor. A Paper Presented at the *Social Science Fourteenth Annual Symposium: The transformation of Middle Eastern urban landscapes: From Modernism to Neoliberalism*. Cairo, Egypt: The American University of Cairo (AUC), May 12–15.

5 Sela

Community and Heritage in Jordan

Maria Elena Ronza

5.1 Introduction

'Sela for Training and Protection of Heritage' (henceforth, Sela) was established in 2015 by the author (currently president of the company) and four colleagues, Eman Abdassalam Ahmad Almousah, Shaker Alfuqara and Bassam Alfakeer, with the stated aim of actively engaging Jordanian host communities in the management, protection and narration of their heritage. Based in Wadi Musa, Sela is the first company of its kind in Jordan and remains unique. At the time of its establishment, it represented a major step in building long-term economic sustainability for host communities through cultural heritage resource management and by pursuing the active involvement of host communities for sustainable heritage preservation. The mere existence of Sela poses serious challenges to an established system that tends to marginalise and exclude host communities from the management of their own heritage.

This chapter will present Sela's work between 2018 and 2024 and its contributions to the development of a new model for the sustainable preservation of the Jordanian cultural heritage through the involvement of the host communities.

5.2 The Jordanian Context

Traditionally, the involvement of host communities in Jordan in the management and protection of their heritage is limited to seasonal unskilled jobs within archaeological missions and conservation projects (Ronza, 2023; Mickel, 2021). On rare occasions, host communities have had an opportunity to contribute to the narration and discourse around the heritage (Badran et al., 2023). In Jordan, archaeological practices have played a pivotal role in perpetuating colonial dynamics of power well into the twentieth century and have ignited a process of dispossession and alienation of host communities from their own past through a patronising attitude (Meskell, 2020, p. 554).

The Department of Antiquities (DoA) of Jordan was established in 1923, and Jordan's first antiquities law dates to 1934. With the 1988 amendment to the Antiquities Law no. 21, most archaeological sites were defined and fenced, creating the first disconnect with the communities living in the surroundings. Since

DOI: 10.4324/9781003491071-6

the early days of the British mandate, and increasingly during the sixties and the seventies, archaeological and later cultural resource management (CRM)[1] projects in Jordan have been almost exclusively conducted in cooperation with or by foreign missions. This situation has engendered and consolidated over the years a sense of alienation within the communities towards their own heritage. Before the seventies, when most archaeological sites were not fenced, they were an integral part of the life of the surrounding communities because of their historical value and were perceived as part of the urban tissue of villages and towns. Therefore, archaeological and heritage sites were lively places, even if the many ongoing activities often endangered the preservation of the site (e.g. exploitation for building materials). Because of such activities, communities are often considered a threat more than a resource for the sites.

Following the fencing of sites and amplified by the increasing number of foreign projects working in the country, the social perception of archaeological sites began to change, with a focus shifting from their cultural and historical value to their potential touristic value. Archaeological sites became immutable, musealised and isolated compartments in the urban tissue, subjects of scientific investigations and tourist destinations. These changes limited the direct interaction of the communities with the sites. Archaeological sites are, in fact, mainly viewed by their host communities as an economic source for tourist exploitation or as a source of seasonal jobs within archaeological projects. Host communities rarely value archaeological sites as their own heritage, as the representation of their own past and as a cultural resource to learn and showcase their own history, primarily viewing them as a resource to draw tourists to the country (Abu Khafajah, 2014).

For all the reasons listed above and despite the growing need to preserve and manage Jordan's myriad cultural heritage resources, the country's CRM training and employment sector remains poorly defined and largely lacks a legal framework by which professionals and technicians can operate in the sector.

The persistence of the colonial legacy continues to foster the narrative of the 'vulnerable other' (Freire, 2005, p. 45; Abu Khafajah and Miqdadi, 2019), which is embedded in the larger scheme of the so-called donor culture that creates dependency and need for validation (Sharp, 2013). And the concept of 'lucrative non-knowledge' introduced by Mickel (2021) is a paradigm of this status quo with the host communities trapped in the role of beneficiaries with needs to be addressed and problems to be solved and not assets to be capitalised.

5.3 A Comprehensive Approach to Community Engagement and Heritage Preservation

5.3.1 Training and Job Creation

Sela has attempted, and has had success, in subverting the paradigm that relegates host communities to the execution of menial jobs without the possibility of a career within the heritage sector by creating cultural heritage preservation and awareness opportunities that enable host communities to become primary stakeholders.

With the support of the Drosos Foundation, Sela launched the Employment through Heritage Project (EHP) in 2018 in collaboration with the DoA. The EHP has created the possibility for an equal partnership between government/research authorities and local communities operating towards three main results:

1 The introduction of professional and technical job profiles to define the legal framework for the CRM workers.
2 The creation of a technical training programme to equip young technicians to operate in the sector.
3 The provision of fieldwork opportunities to university students and fresh graduates to prepare them for the job market.

Sela worked with the DoA to define professional and technical profiles for the CRM workers and to design training programmes responding to the needs of the field. Through its training courses, Sela transformed a seasonal unskilled labour force into a formal workforce of trained technicians that successfully entered the job market within the newly defined technical profiles. The definition of qualifications facilitated the change in hiring practices with the introduction of written formal contracts in a sector that traditionally privileges short-term unregistered verbal agreements (Corbett and Ronza, 2022). Over 35 national and international CRM projects in 19 communities in Jordan adopted the EHP model, and over 360 individuals completed their training between 2018 and 2022 (Figure 5.1).

Despite the success of EHP in creating job opportunities, the CRM sector in Jordan remains mostly seasonal. Most employment opportunities are still provided within foreign-led projects and by the Jordanian government through seasonal socio-economic initiatives. Figure 5.1 shows the employment rate of Sela's trainees within one year after completing at least one cycle of training. The average rate of 71% over the four years of implementation of EHP demonstrated the potential for this labour market to become a sustainable industry.

Building on the successes of EHP, in 2022 Sela launched 'Jordanian Archaeology as a Sustainable Industry (JASI)'. This project aims at piloting a multidisciplinary training approach to diversify and enrich the training experience by including soft

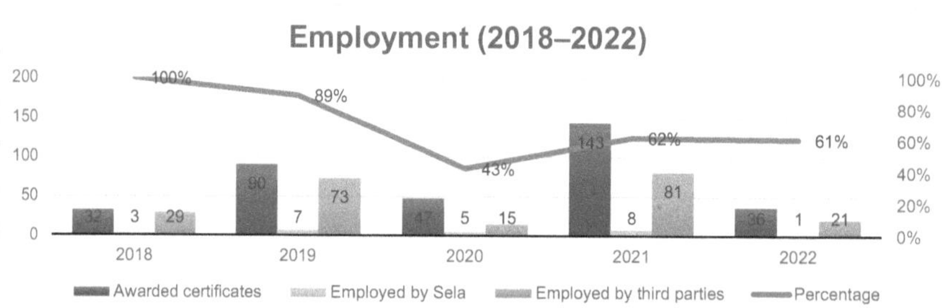

Figure 5.1 Employment rate of Sela's certified trainees from 2018 to 2022.

Figure 5.2 Mohammad Albdoul, one of Sela's trainers, giving a theoretical training session on archaeological documentation at the Great Temple in Petra. © Sela.

skills that are crucial to achieve competitiveness within the job market. The project is designed to expand the results of EHP and further promote public/private cooperation in the management of archaeological cultural resources.

The proposed model has proved its efficiency within the first year of implementation. In May 2023, Sela started a long-term collaboration with the Petra Development and Tourism Region Authority (PDTRA) to support the training and employment of the young members of the host communities living around the World Heritage Site (WHS) of Petra to secure their involvement in the routine maintenance of the site (Ronza, 2022). Within the first year of this collaboration, over 200 young people from the six communities of the Petra region have received technical training and have been hired with long-term contracts to implement the maintenance of the site (Figure 5.2).

Besides the fieldwork, Sela promotes a three-tiered participatory approach designed to involve the entire community from the macroscopic level to the individuals:

1 The civil society is involved through community-oriented activities and awareness campaigns to enhance the sense of ownership towards the heritage and raise awareness on the importance of its protection.
2 The community of practice[2] through interdisciplinary training programmes that include the exchange of experiences and interaction between community members who share common values and interests (e.g. cultural heritage preservation).

3 The individuals through training programmes, job creation and vocational guidance.

Young professionals, new graduates, technicians and school dropouts have been involved in the management and conservation of the architectural heritage of Jordan through specialised training that equips them with marketable skills. Fieldwork opportunities have been granted to university students within an international work environment to enhance their ability to work in cross-cultural, multidisciplinary teams and gain practical experience that might provide guidance for their professional and scientific paths and marketable work experience. Finally, several heritage-themed activities have been designed and proposed to school-age children and their families to positively impact their perception of the CRM professions as an alternative to tourism opportunities being the only value of archaeological sites.

Central to Sela's approach is the recognition of the indispensable role of local stakeholders in the sustainable preservation of cultural heritage. From adults and children to educators and students to community leaders, Sela actively involves a diverse array of voices in its initiatives.

5.3.2 Raising Awareness Initiatives

Engaging the host communities in the discourse about the heritage and bringing their voices into the narration serve to empower and enable individuals to take ownership of their cultural heritage and instil a sense of pride in their role as stewards of the past, recognising it as an integral part of their identity and collective memory. Sela has undertaken concerted efforts to enhance the experience of visitors at archaeological sites and has collaborated with national and international partners to install informative panels at key sites. These panels offer insights into the historical significance of the cultural heritage and the needs for its conservation to local and international visitors.

Sela has also engaged the younger generation through targeted educational programmes. The 'Little Archaeologist' programme stands as a testament to Sela's commitment to nurturing an appreciation for cultural heritage from a young age. Through hands-on activities such as mosaic making and pottery workshops, children are introduced to history in a playful way. These immersive activities ignite children's curiosity and imagination, laying the foundation for a lifelong journey of discovery and appreciation for cultural heritage (Figure 5.3). Sela delivers these activities at schools, social centres, museums and during children's seasonal camps. These camps, targeting children between 7 and 16 years old, offer participants the opportunity to delve deeper into the world of archaeology and heritage preservation. Through a combination of educational workshops, guided tours and heritage-themed activities, children are educated about their cultural heritage. The camps serve as platforms for community engagement, bringing together children from diverse backgrounds and from different communities to share in the joy of discovery and exploration.

Figure 5.3 One of the participants of the winter camp at Faynan proudly showing his creation in copper. © Sela.

Beyond their educational and recreational value, Sela's camps play a crucial role in instilling a sense of responsibility and stewardship for the future generations of cultural ambassadors and advocates. Sela has also designed and produced several educational tools for Jordanian public museums, such as stratigraphic boxes for five public museums, two cartoon videos for the museum in Faynan and a game for the Petra Museum (Figure 5.4).

In essence, Sela's cultural heritage strategy embodies a holistic approach to community engagement and preservation. By empowering host communities, enhancing visitor experiences and nurturing the next generation of stewards, Sela contributes to the safeguard of Jordan's cultural heritage but also lays the foundation for a future where heritage preservation is a shared responsibility.

5.4 Al-Raqeem: Fostering Sustainable Conservation and Economic Development

In cultural heritage preservation, collaboration is key to amplifying impact and ensuring long-term sustainability. In 2020, the author and Ahmad Almousah established another company, Al-Raqeem, with the aim of supporting Sela's mission. The main objective is the diversification of the income stream and the creation of a sustainable income for Sela by providing archaeological and heritage conservation services and therefore employment opportunities for Sela's trainees.

Figure 5.4 The author leading a visit to the Amman Citadel Museum to play with the stratigraphy box. © Sela.

Al-Raqeem was also established to provide a mechanism to develop local community businesses which can offer services to archaeological projects or sustainable heritage-themed products.

Recognising the importance of economic sustainability in heritage preservation efforts, Al-Raqeem seeks to empower local communities by creating viable livelihood opportunities centred around cultural heritage. By providing training and support to entrepreneurs interested in heritage-related businesses, it enables communities to become active stakeholders in heritage conservation. Through the establishment of sustainable enterprises such as handicraft shops, cultural tours and artisanal workshops, Al-Raqeem fosters economic development and empowerment at the grassroots level.

Al-Raqeem owns a registered trademark for its initiatives, Entimaa', which involves producing and selling heritage-related products and experiences. Currently, Al-Raqeem, through Entimaa', is marketing several products, including locally handmade products, archaeology-themed toys and food products. Al-Raqeem donates a minimum of 25% of its profit for heritage conservation.

In 2021, Al-Raqeem/Sela started a collaboration with the Jordanian writer, Rasha Dababneh, and produced a series of storybooks targeting 6–8-year-old children to engage them with their heritage. The initiative received a lot of attention, and several projects requested to produce a book about their site. Currently, six books have been published about the Amman Citadel, Tell Dhiban, Tell Hisban,

Faynan and Petra. The books are written with the input of the children living around the sites where the books are set. Another innovative initiative pioneered by Al-Raqeem is 'Once Upon a Taste…,' a culinary experience that celebrates Jordan's rich gastronomic heritage.

5.5 Sela's Impact

Within its main mission of actively engaging Jordanians in the management of their heritage, Sela is committed to the achievement of the Sustainable Development Goals (SDGs). Its projects and initiatives have contributed to 9 of the 17 SDGs, as follows:

Goal 1 – No Poverty: Sela has created over 500 short- or long-term job opportunities for young Jordanians since 2015. According to the Department of Statistics, the unemployment rate in Jordan is 22.3% in 2023. Since its foundation in 2015, through its training programmes, Sela has equipped over 700 young people with marketable skills and over 70% of the trainees found a job opportunity within one year from the end of their training. Sela's website hosts an open-access database of trained Jordanian technicians to help former trainees to access the job market.

Goal 4 – Quality Education: Sela offers a series of vocational training programmes in coordination with and in response to the needs of the major stakeholders in the sector, e.g. the DoA and the PDTRA and several national and international archaeological missions. Within those programmes, Sela has trained over 700 young individuals. The majority of the programmes proposed by Sela are designed to grant inclusivity, especially to those who have had no access to formal education. Since the training programmes are designed and implemented in collaboration with the sector's stakeholders and major employers, they provide skills that are immediately marketable within the CRM projects, but also across other fields within the region.

Goal 5 – Gender Equality: The female unemployment rate in Jordan in 2023 is 30.9%.[3] Several studies link women's unemployment to increased domestic violence (Oweis et al., 2009, pp. 69–76; Akilova et al., 2014, p. 67) and early marriages (Gebel and Heyne, 2016; Hikmat, 2017). By pursuing a gender-blind, skill-based policy, Sela offers the opportunity to change and normalise the image of working women in patriarchal societies, especially in worldwide male-dominated sectors, such as archaeology and conservation. With over 35% of the job opportunities given to females, Sela contributes to empower women to advocate for themselves.

Goal 8 – Decent Work and Economic Growth: According to the 2022 World Bank report, 59% of the total employment in Jordan is informal, as is 75% of private sector employment (Islam et al., 2022, p. 10). With the creation of formal job opportunities for young women and men, Sela's projects fostered decent employment conditions and a safe and inclusive work environment.

Goal 10 – Reduced Inequalities: Sela's projects provide opportunities for marginalised communities in Jordan to actively participate in decision-making processes related to heritage conservation and management. Such inclusion empowers communities' members to advocate for themselves and contributes to reducing inequality by voicing the rights of the civil society. With JASI, Sela worked extensively to raise awareness and educate community members about cultural heritage through heritage-themed activities promoting cultural understanding, tolerance and empathy, which helped challenge stereotypes and prejudices and promoted equality and respect between different cultural groups.

Goal 11 – Sustainable Cities and Communities: By promoting a circular economy system through prioritising local products, traditional techniques and building local capacity, Sela aims to provide long-term sustainability to cultural heritage. In addition to that, by creating jobs within the CRM sector, Sela contributes to the sustainability of the communities both financially and in terms of well-being through the lessons of resilience that history can teach.

Goal 12 – Responsible Consumption and Production: Through its projects and training programmes, Sela promotes an approach that prioritises locally produced conservation materials. Sela is conducting several research projects to develop locally produced environmentally friendly conservation products.

Goal 16 – Peace, Justice and Strong Institution: Sela's projects seek a strong collaboration between government institutions and civil society. These projects empower the communities by involving them in decision-making and implementation processes. Such empowerment strengthens local institutions as community members take an active role in shaping the future of their environment and society.

Goal 17 – Partnerships for the Goals: Both the 'Employment through Heritage' and the 'Jordanian Archaeology as a Sustainable Industry' projects are model examples of private/public collaboration with the active engagement of the civil society. They contribute to building strong relationships with government institutions through joint capacity-building programmes.

5.6 Reflections

The establishment of Sela marks a significant milestone in the CRM sector in Jordan, demonstrating the positive contributions that local communities can make at a national level. While Sela has achieved commendable success, there exists a crucial need for established cultural heritage stakeholders to create a legal and professional environment that recognises local communities as equal partners in the safeguarding of heritage. Sela has emerged as a key player in connecting diverse stakeholders and initiating collaborative projects at the local level. It has brought fresh perspectives to CRM and public archaeology in Jordan, acting as a liaison between governmental authorities, Jordanian civil society organisations and international institutions. Through these efforts, Sela has facilitated long-term collaborations that have significantly contributed to heritage preservation. However, the sustainability of Sela's community engagement model hinges on a transformative shift in the existing paradigm that challenges the conventional relegation of local communities to peripheral roles in heritage management. Sela seeks to dismantle this outdated approach by championing a more sustainable model that recognises the inherent value of local knowledge, skills and perspectives. Sela advocates for a more inclusive and sustainable approach that positions local communities as active and equal partners in the preservation of Jordan's rich cultural heritage.

Sela is piloting a model that aims to formalise collaborative approaches, fostering synergies among diverse entities, including public and private stakeholders at national and international levels. Sela's initiatives seek to bridge the gap between local and global perspectives, establishing multidisciplinary skilled units capable of overseeing heritage projects from inception to completion. By creating a framework for collaboration and skills development, Sela endeavours to empower local communities, ensuring their active involvement in the entire lifecycle of heritage projects. This holistic strategy is essential for cultivating a sense of ownership and responsibility among local stakeholders, contributing to the long-term sustainability of heritage preservation efforts. Sela's approach recognises the interconnectedness of heritage, community well-being and sustainable development.

Looking ahead, continued collaboration and innovation will be essential in safeguarding Jordan's cultural legacy for future generations. In a rapidly changing world, where the pressures of urbanisation, globalisation and climate change threaten to erode cultural heritage, the work of organisations like Sela and Al-Raqeem has never been more critical. By remaining steadfast in their commitment to community engagement, education and sustainable tourism, they can ensure that the cultural heritage continues to serve as a source of inspiration, pride and prosperity for generations to come.

Acknowledgements

I extend my heartfelt gratitude to the Drosos Foundation for their generous support over the years, which has played a pivotal role in advancing Sela's mission. Additionally, my sincere appreciation goes to the PDTRA for their invaluable financial and logistical assistance, which has significantly contributed to the success of Sela's projects. I am deeply thankful to the DoA for their enduring partnership and expertise. The longstanding collaboration with DoA has been instrumental in shaping and implementing Sela's projects. Special recognition is due to Professor Oysten Labianca for his mentorship and unwavering support. His guidance has been instrumental in Sela's efforts to actively engage communities in the management of their heritage, and I am truly grateful for his expertise and encouragement. I would also like to express our appreciation to the project directors who have believed in and supported our vision, like Professor Mithen from the University of Reading, Professor Routledge from Liverpool University, Professor Shatha Abu Khafajah from the Hashemite

University, Dr. Fatma Marii and Dr. Arwa Badran for their collaborative efforts and contributions to our projects. Finally, I would like to extend my acknowledgements to those who supported Sela's mission for their dedication and commitment to the preservation and management of cultural heritage.

Notes

1 Throughout this document, CRM indicates the practice of managing cultural resources, which demands a multidisciplinary approach. For the origin and definition of the term, see King (2020). www.encyclopedia.com/social-sciences/applied-and-social-sciences-magazines/cultural-resource-management (last visited May 2024). www.jstor.org/stable/20210052?seq=1#page_scan_tab_contents (last visited May 2024).

2 'A community of practice is a group of people who share a concern or a passion for something they do and learn how to do it better as they interact regularly.' This definition reflects the fundamentally social nature of human learning. It is very broad. It applies to a street gang, whose members learn how to survive in a hostile world, as well as a group of engineers who learn how to design better devices or a group of civil servants who seek to improve service to citizens. http://wenger-trayner.com/resources/what-is-a-community-of-practice/ (last visited May 2024).

3 That the female unemployment rate in Jordan in 2023 was 30.9% is the figure given by the World Bank in Jordan within its report for Jordan available at www.worldbank.org/en/country/jordan/overview#:~:text=Jordan%27s%20labor%20force%20participation%20rate,%25)%20being%20the%20most%20affected. (Last visited May 2024).

References

Abu-Khafajah, S. (2014). 'They Are Hiding It … Why Do They Hide It? From Whom, and For Whom? Community Heritage at Work in the Post-Colonial Context of Jordan', in: S. Thomas, J. Lea, A. Nissinaho, A. Pupio, A. McGill, A. Badran, B. Camargo, C. MacDonald, C. Alegria, D. Saraç, G. Jorayev, J. Moe, J. Clark, J. Debert, L. Aranda, M. Brody, M. Corbishley, N. Mazzia, S. Abu-Khafajah, S. Lampe, T. Moussouri, T. Kador, T. Soininen and Salerno, V. (eds.) *Public Participation in Archaeology*. Woodbridge: Boydell and Brewer, pp. 149–160.

Abu-Khafajah, S. and Miqdadi, R. (2019). Prejudice, Military Intelligence, and Neoliberalism: Examining the Local within Archaeology and Heritage Practices in Jordan, *Contemporary Levant*, 4(2), pp. 92–106.

Akilova, M. and Marti, Y.M. (2014). What Is the Effect of Women's Financial Empowerment on Intimate Partner Violence in Jordan?, *Glob Social Welfare*, 1, pp. 65–74.

Atalay, S. (2010). We Don't Talk about Çatalhöyük, We Live It: Sustainable Archaeological Practice through Community-Based Participatory Research, *World Archaeology*, 42(3), pp. 418–29.

Badran, A., Abu-Khafajah, S., Ronza, M.E., Skeates, R., Wilkinson, R. and Marii, F. (2023). 'Collaboration and Multivocality in Heritage and Museum Practice: Lessons from Jordan', in: R. Bonnie, M. Lorenzon and S. Thomas (eds.) *Living Communities and Their Archaeologies in the Middle East*. Helsinki: Helsinki University Press, pp. 153–190.

Corbett, G.J. and Ronza, M.E. (2022). 'Making Social Engagement Sustainable: Insights from the Temple of the Winged Lions Cultural Resource Management Initiative in Petra, Jordan', in: A. Badran, S. Abu-Khafajah and S. Elliott (eds.) *Community Heritage in the Arab World: Values and Practices*. New York: Springer, pp. 49–64.

Freire, P. (2005). *Pedagogy of the Oppressed.* New York: The Continuum International Publishing.

Gebel, M. and Heyne, S. (2016). Delayed Transitions in Times of Increasing Uncertainty: School-to-Work Transition and the Delay of First Marriage in Jordan, *Research in Social Stratification and Mobility,* 46*(Part A)*, pp. 61–72.

Hikmat, R. (2017). *A Study on Child Marriage in Jordan.* Jordan: The Higher Population Council.

Islam, A., Moosa, D. and Saliola, F. (2022). *Jobs Undone: Reshaping the Role of Governments Toward Markets and Workers in the Middle East and North Africa.* Washington: International Bank for Reconstruction and Development/The World Bank.

King, T.F. (2020). *Cultural Resource Management: A Collaborative primer for Archaeologists*. New York: Berghahn Books.

LaBianca, Ø. and Ronza, M.E. (2022). Site Custody Activism in Jordan: Introducing Sela and Alraqeem, *Levant*, 54(3), pp. 444–450.

Meskell, L. (2020). Imperialism, Internationalism, and Archaeology in the Un/Making of the Middle East, *American Anthropologist*, 122, pp. 554–567.

Mickel, A. (2021). *Why Those Who Shovel Are Silent. A History of Local Archaeological Knowledge and Labor*. Colorado: University Press of Colorado.

Oweis, A., Gharaibeh, M., Al-Natour, A. and Froelicher, E. (2009). Violence Against Women: Unveiling the Suffering of Women with a Low Income in Jordan, *Journal of Transcultural Nursing*, 20(1), pp. 69–76.

Ronza, M.E. (2022). The impact of Sela for Training and Protection of Heritage on the Jordanian CRM sector. *Archeostorie. Journal of Public Archaeology*, 5, pp. 1–22 (pre-print). DOI: https://doi.org/10.23821/2022_3c/

Ronza, M. E. (2023). 'Do Communities Have a Role in Community Archaeology in Jordan?', in: R. Bonnie, M. Lorenzon and S. Thomas (eds.). *Living Communities and Their Archaeologies in the Middle East*, Helsinki: Helsinki University Press, pp. 11–32. DOI: https://doi.org/10.33134/HUP-19-2

Sharp, J. (2013). 'Africa's Colonial Present: Development, Violence, and Postcolonial Security', in: G. Huggan (ed.) *The Oxford Handbook of Postcolonial Studies*. Online: Oxford Academic, pp. 235–252.

6 The Land of Nineveh Archaeological Project

Cultural Heritage Protection and Enhancement, Community Engagement, and Sustainable Tourism in the Kurdistan Region of Iraq

Daniele Morandi Bonacossi, Francesca Simi, and Bekas Jamaluddin Hasan

6.1 Introduction[1]

Although the relationship between heritage and development has long been considered (Labadi, 2022), academics concerned with heritage have recently become increasingly interested in investigating the role of heritage in local sustainable development (e.g. Giliberto and Labadi, 2022; Labadi *et al.*, 2021; Baillie and Sørensen, 2021; Hampton, 2005; Cernea, 2001).

Notwithstanding its potential negative socio-cultural and ecological impacts (Comer and Willems, 2011; Timothy and Nyaupane, 2009, pp. 59–65; Wall and Mathieson, 2006), heritage tourism can serve as an essential means of conserving cultural heritage. As we will illustrate by our case study of 'Sennacherib's Archaeological Park', this can be achieved by funding conservation and monitoring efforts and providing a rationale for establishing parks and areas of protected heritage (Timothy, 2014, p. 32).

The shift from a sole conservation and/or research-focused approach to archaeological sites to one that includes development goals is already integrated into the postcolonial and decolonial discourse that now permeates the heritage disciplines (Labadi, 2019; Labadi, 2022), particularly archaeology (Gould and Burtenshaw, 2019). Archaeologists are increasingly called to contribute to contemporary issues at the local and global levels (e.g. Matsuda and Okamura, 2011).

In a still ongoing epistemological and ethical transformation of the discipline, archaeology-based tourism or archaeo-tourism is progressively recognised as a manifestation of public archaeology (Timothy and Tahan, 2020; Burtenshaw and Palmer, 2014; Lenik, 2013; Newell, 2008). Notwithstanding some averseness in the past among heritage specialists to becoming involved in tourism (Burtenshaw, 2014; Deacon, 2006), there is a growing urgency to bridge the gap between research and application, archaeology and the public.

DOI: 10.4324/9781003491071-7

In post-conflict situations, sadly widespread in West Asia, where heritage is often used for national, political, economic and religious agendas, projects that aim to create a symbiosis between heritage protection, archaeological research, and sustainable development through tourism are encountering several peculiar challenges (Hall and Seyfi, 2021, p. 34). In our study area of the Kurdistan region of Iraq (KRI), the relationship between heritage protection and economic development continues to be a cause of tension, as is often the case elsewhere. Until recently, the heritage sector in Kurdistan had been weak, with limited governance and little presence in discussions about development agendas. During the last decade, however, a new wave of archaeological projects coupled with the Kurdistan Regional Government's (KRG) interest in diversifying its economy are rapidly changing this situation.

In this chapter, we will attempt to reflect on which issues relating to archaeology, conservation, and sustainability are specifically relevant to the KRI. Furthermore, we will consider the challenges faced by post-colonial heritage/archaeological practices intending to promote democracy, human rights, and equal access to resources and contribute to the ideals of 'development' as stated in the UN Sustainable Development Goals. In this sense, as we will explain, the role of local communities and local stakeholders becomes pivotal.

The starting point of our concerns is represented by the 'Land of Nineveh Archaeological Project' (LoNAP) that the University of Udine has been conducting since 2012 in the Duhok Governorate. LoNAP is an interdisciplinary landscape archaeology project that aims to understand the formation and transformation of the cultural and natural landscape of a strategic region of northern Mesopotamia, an area that embraces large parts of the governorates of Ninawa and Dohuk. The archaeology extends from the Palaeolithic to the Islamic period, and the project seeks to secure its protection and valorisation using innovative strategies (Morandi Bonacossi and Iamoni, 2015).

LoNAP is based on an archaeological survey of approximately 3,000 km^2 and the excavation of the sites of Tell Gomel and Faida, aiming to explore the organisation of the Assyrian Empire (c. 1350–600 BCE) on the ground and its impact on societies and landscape in its core region. Assyria's imperial success story – which led to the empire extending from the Gulf region and western Iran to Iraq, southeastern Turkey, the Levant, and Egypt – was made possible by its ability to intervene profoundly in its core region and periphery with practices that interfered with local societies, economies, and landscapes (Morandi Bonacossi, 2018a-b; Düring, 2020). An example of Assyria's impact is the impressive regional irrigation system that was built in the hinterland of the last imperial capital, Nineveh. This has been one of the main focuses of LoNAP's work in the Duhok region.

6.2 Heritage Protection and Tourism in the Kurdistan Region of Iraq

After the fall of the Baathist regime in 2003, the KRI experienced a rapid economic, cultural, and social regeneration that remains ongoing. The stability and security of the region, coupled with the welcoming attitude of the country in general, and particularly of the KRG General Directorate of Antiquities and Heritage, contributed to the establishment of dozens of new archaeological projects in the region (Kopanias *et al.*, 2015; Kopanias and MacGinnis, 2016; Orazi, 2019; Ur, 2017). This renaissance of archaeological exploration in north-eastern Iraq enabled the mapping of thousands of previously unknown archaeological sites, contributing enormously to our understanding of the patterns of settlement and land use from earliest prehistory to the modern era (Ur, 2017). At the same time, these projects focused on recording, protecting, and enhancing cultural heritage, as well as monitoring threats to its preservation from uncontrolled development and vandalism (Mardas, 2017; Morandi Bonacossi and Tonghini, 2018; Morandi Bonacossi, 2021). The urgent need to protect an extremely rich and diverse heritage soon became apparent: rapid demographic growth and urbanisation (Stansfield, 2003; Sabr, 2014; Jarah *et al.*, 2019) are having an enormous negative impact on the archaeological sites and landscapes. The economic surge, administrative autonomy, and other political developments are at the roots of the region's population growth; according to the Kurdistan Region Statistics Office, the population increased by 41 per cent between 2013 and 2023, from 4.66 to 6.56 million persons (IOM, 2024; KRSO, 2024; Osman, 2021). Overall, the quest for modernisation and development of infrastructure, industrial and agricultural activities, and housing construction has become inexorable.

In practice, archaeological mapping and recording have become a race against time. The numerous archaeological projects active on KRI territory and the pooling of newly collected data with the General and Local Directorates of Antiquities and Heritage of the KRI have provided a new tool for heritage protection. Besides the more traditional threats to cultural heritage, such as vandalism, illegal excavations, and the clandestine trade in antiquities, those in the KRI include the intensification of agricultural activities, urban sprawl, industrialisation (including oil and gas extraction plants, new factories, illegal refineries), the creation of extensive road infrastructures, and the uncontrolled exploitation of mineral resources, notably gravel extraction from riverbeds for cement production. The region's most iconic and monumental sites, such as the rock reliefs and historic buildings, have not been spared from unthinking and often unintentional damage or destruction. Several international stakeholders have been or are currently active in the KRI, and more widely in Iraq, to support the protection and valorisation of archaeological sites. These include the ALIPH Foundation (the International Alliance for the Protection of Heritage in Conflict Areas), Gerda Henkel Stiftung, World Monuments Fund, UNESCO, J.M. Kaplan Fund, Italian Agency for Development Cooperation, and the British Council.

In the last fifteen years, the region's stability and relative security have also been significant. Tourism has boomed with over four million tourists visiting the region in 2021, six million in 2022 and over seven million in 2023. Considering this positive trend, the KRG Tourism Board has set an ambitious goal of attracting 20 million tourists annually by 2030. Among the many initiatives to support this new economic sector, the KRG launched an electronic visa application system in 2023 to facilitate international tourist traffic into the Kurdish region.

Over the past two decades, Kurdistan's economy has been heavily reliant on the oil and gas sector, which has been the key driver of the region's development. In line with many other West Asian countries (Seyfi and Hall, 2021, p. 43; Timothy, 2019), Kurdish authorities are trying to diversify the KRI economy, placing tourism into its new economic strategy. National (especially from the southern Iraqi provinces) and international tourism (especially from Iran and Turkey, but also from Gulf Countries, Asia, and Europe) represent a key element for the economic, social, and cultural growth of the region.

The promotion of tourism was already a top priority of the 'Kurdistan Region of Iraq 2020, A Vision for the Future', developed and published in 2013 by the KRG Ministry of Planning. More recently, the diversification of the economy through the promotion of the tourism industry remains one of the strategic points of the 'KRI 2030 vision' drafted by the KRG Ministry of Planning with the support of the United Nations Development Programme (UNDP). In the 'Reform Roadmap for the Tourism Sector in Kurdistan Region (2022)',[2] developed under the leadership of the UNDP in Iraq together with the KRG General Board of Tourism, heritage sites are recognised as critical assets of the tourism sector. The reform proposes to strengthen the management of tourism policies, including the updating of legislation, classification, licensing, educational, tourism, and statistical systems and the development of international relations. The KRI has the potential to attract numerous tourists due to several factors, including its stunning natural and archaeological landscapes and the high quality of the heritage sites preserved in its territory. The new law for the Management and Protection of Antiquities and Heritage (KRG Law No. 5 of 2021)[3] also pinpoints in Article 41 the importance of 'making archaeological and heritage sites touristic, provided that their original features are protected'.

Several international and local projects are now working to support the growth of eco- and archaeo-tourism for sustainable development in the KRI, among which are the Erbil Citadel Conservation and Rehabilitation Project (conducted by the High Commission for the Erbil Citadel Revitalization and funded by several international and local institutions including UNESCO, the US Embassy, and the World Monuments Fund) and, on a smaller scale, the Charsteen Conservation Project (Duhok Directorate of Antiquities and Heritage (DDAH), with the support of UNESCO, EU, and ILO). The 'Central Zagros Archaeological Project' (University of Reading, UK)[4] and our 'LoNAP' (University of Udine, IT)[5] are two archaeological projects engaged in the protection and valorisation of the KRI cultural heritage while promoting new opportunities for local sustainable economic development through tourism.

Notwithstanding that Iraq, comprising the larger portion of ancient Mesopotamia, is known to have been the 'cradle of civilisation', the country has only six World Heritage Sites (WHSs) (UNESCO, 2024a): Hatra since 1985, Ashur since 2003, Samarra since 2007, the Erbil Citadel since 2014, the Ahwar of Southern Iraq since 2016, and Babylon since 2019. The first three of these sites are also in the List of World Heritage in Danger. Among these, only one, the Erbil Citadel, is in the KRI. In Iraq's World Heritage Tentative List (UNESCO, 2024b), there are another 14 sites of which three are in the KRI (Amedy city since 2011, Lalish Temple since 2020, and the Bestansur site since 2017). The inclusion of other new archaeological sites in the World Heritage List (WHL) would help to expand archaeological tourism in the region. For this reason, LoNAP is liaising with the General Directorate of Antiquities and Heritage of the KRG and the State Board of Antiquities and Heritage of Iraq in the submission of an application to UNESCO to include Sennacherib's Irrigation Network in the World Heritage Tentative List as a first step to its recognition as a WHS.

6.3 Sennacherib's Northern Assyrian Irrigation System

The core region of the Assyrian Empire between the Upper Tigris and Euphrates valleys of Iraq and Syria (Figure 6.1) is part of the stable dry-farming zone of Upper Mesopotamia (Wilkinson, 2004, pp. 42–43). Here, water supply from rainfall is reliable, and cereals can be predictably and extensively cultivated using a rain-fed system. However, recurrent episodes of climate variability – resulting in harsh drought events as mentioned by textual sources and now documented by isotope analysis of speleothems (Sinha *et al.*, 2019) – could seriously endanger the success of the crops. For this reason, the widespread intensification of rural settlement in the Assyrian countryside and the urban and demographic development of its capitals and provincial centres had involved the reorganisation of Assyria's natural hydrology and the creation of centrally planned hydraulic systems (Morandi Bonacossi, 2018a–b; Morandi Bonacossi and Qasim, 2022). Massive canal networks engineered and redirected rivers, wadis, and karst springs into channels that fed cities and irrigated the countryside (Morandi Bonacossi, 2018b; Morandi Bonacossi and Qasim, 2022; Ur *et al.,* 2013, pp. 105–106; Ur et al., 2021, pp. 24–26; Ur and Reade, 2015).

In the hinterland of his new capital, Nineveh, the Assyrian ruler Sennacherib (704–681 BCE) accomplished the construction of the most ambitious and extensive irrigation network ever built in Assyria (Figure 6.2), now referred to as the Northern Assyrian Irrigation System. For over 240 km, a branched network of irrigation canals was excavated in the earth of the eastern Tigris alluvial plains or hewn into the limestone bedrock of the Zagros hilly flanks, consisting of canalised seasonal watercourses, embankments, stone aqueducts, reservoirs, and weirs. This irrigated the countryside behind Nineveh bestriding the modern governorates of Duhok and Ninawa. It also brought water to the fields around the ancient capital, to the palace and to the royal parks built by the king within Nineveh's upper town.

The Northern Assyrian Irrigation System was created by Sennacherib in four stages. The third and fourth stages of the system fall within LoNAP's research area. The third stage consisted of four canals. The plain immediately south of the modern city of Duhok was irrigated by a pair of parallel, massive cross-watershed canals close to Maltai and a smaller canal dug into the rock close to the modern town of Faida (Morandi Bonacossi and Qasim, 2022). A third gigantic canal carried water from the Wadi Bandawai to a tributary of the Wadi Al Milh and to a fourth large

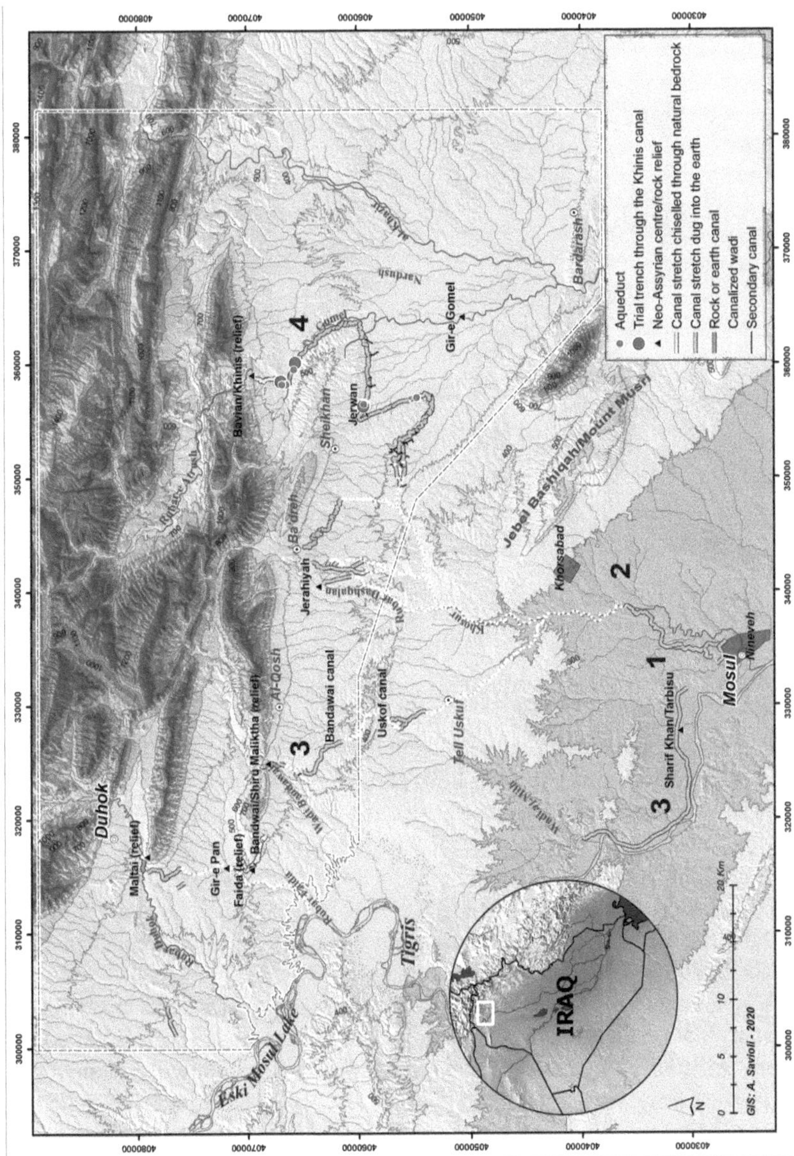

Figure 6.1 The Northern Assyrian Irrigation System and the sites involved in the Sennacherib's Archaeological Park. © LoNAP.

Figure 6.2 The Khinis Archaeological Complex. Photo: F. Simi, © LoNAP.

canal stretch close to Tell Uskof (Morandi Bonacossi, 2018b). The latter brought the water to a tributary of the Khosr (Ur, 2005, p. 332).

Monumental commemorative rock reliefs at Maltai and Faida portray the Assyrian ruler represented twice, at both ends of each panel, in front of the cult statues of seven divinities in the shape of striding animals or mythological creatures standing on pedestals. Four panels are known from Maltai, while the archaeological exploration of the Faida canal (Figure 6.3) by a joint Kurdish–Italian Project from 2019 to 2022 revealed thirteen reliefs (Morandi Bonacossi and Qasim, 2022). At Shiru Maliktha, at the head of the Bandawai canal, a rock relief represented the king beneath divine symbols (Morandi Bonacossi, 2018b).

The Khinis canal system was the fourth and last stage of Sennacherib's irrigation programme. At around 690 BCE, the waters of the River Gomel were diverted into a tributary of the Khosr by means of a 55-km-long canal that started at Khinis at the exit of the river from a narrow mountain gorge (Jacobsen and Lloyd, 1935, pp. 44–49; Morandi Bonacossi, 2018b-c). To celebrate his programme of centralised landscape planning, Sennacherib ordered the execution of monumental commemorative reliefs on the cliff face. A monolith sculpted with bas-reliefs marked the canal head, where the water of the River Gomel was redirected into the canal by means of a stone weir. Towering over the canal, the imposing 'Large Panel', the

Figure 6.3 One of the thirteen reliefs discovered at Faida. Photo: D. Tagen, © LoNAP.

'Rider Relief', and twelve rock niches with bas-reliefs depicting Sennacherib's image beneath divine symbols were carved in the rock. Three of the bas-reliefs contained the so-called Bavian Inscription of c. 688 BCE, which commemorated the excavation and inauguration of the canal that the king called 'Sennacherib's canal' (Grayson and Novotny, 2014, pp. 310–317).

The Khinis system is unique among Sennacherib's canals not only due to its monumental canal head and branching, but also because it preserves important royal building inscriptions. Besides the 'Bavian Inscription', a series of shorter inscriptions on the aqueduct at Jerwan – the most monumental of the five aqueducts existing along the Khinis canal – celebrate the construction of this massive infrastructure project (Jacobsen and Lloyd, 1935, pp. 20–21; Fales and Del Fabbro, 2014).

The construction of large infrastructure such as Sennacherib's canal network undoubtedly had a major impact on the Assyrian economy, ensuring a continuous supply of water to Nineveh and significantly increasing the production capacity of its agricultural hinterland. The impressive rock reliefs at Khinis, Maltai, Faida, and Shiru Maliktha and the aqueduct of Jerwan represent a unique and coherent complex of monumental archaeological sites, which are distributed throughout a vast territory, often in isolated places. As such, they represent – together with the entire Northern Assyrian Irrigation System in which they are embedded – testimony to the extraordinary ability of ancient empires to impose change upon the landscape while respecting its characteristics and morphology: Sennacherib's irrigation system was an ecologically compatible landscape design, providing an efficient and sustainable irrigation project in a territory subject to increasing aridification.

6.4 The Land of Nineveh Archaeological Project (LoNAP)

In 2015, LoNAP started a project aimed at recording, protecting, restoring, and enhancing this unique cultural heritage through the creation of the first archaeological park in the region. Among its main objectives, 'Sennacherib's Archaeological Park' has the development of the scientific, technical, and cultural knowledge of local operators by training courses and work experience and of sustainable tourism that will strongly foster the socio-economic development of the country. To these ends, with the financial support of the Italian Agency for Development Cooperation (AICS), between 2015 and 2020, the entire Northern Assyrian Irrigation System was surveyed and the rock reliefs of Maltai and Khinis and the Jerwan aqueduct were recorded using state-of-the-art techniques, including laser scanning, 3D modelling, and unmanned aerial vehicle (UAV) technology.

Expert stone conservators drafted a report on the state of preservation of all the rock reliefs associated with the irrigation network and a project for their conservation. Training courses on archaeological method and theory, archaeological illustration, bioarchaeology, geoarchaeology, and basic conservation treatment of artefacts were organised for the personnel of the Duhok, Zakho, and Akre Directorates of Antiquities. A conservation laboratory with the basic equipment for first aid conservation was donated to the Duhok Archaeological Museum, and its personnel was trained on-the-job, giving conservation treatment to more than hundred objects from the museum's collections. Lastly, a team of architects with long experience in the study, conservation, and enhancement of architectural heritage drafted a proposal for the creation of the archaeological park and the inclusion of the Assyrian rock reliefs, aqueducts, and canals in the UNESCO World Heritage List. The results of this endeavour were published in a monograph (Orazi, 2019) and made available to a wide audience through a dedicated website.

The 2019 to 2022 excavation of the rock-cut Faida canal, which has thirteen sculpted panels carved along its bank, sparked great attention in the scientific community and led to the inclusion of Faida in the archaeological park project (Morandi Bonacossi and Qasim, 2022). With financial support from the ALIPH and Gerda Henkel Foundation, a specific pilot project dubbed 'Faida Salvage Project' was launched in collaboration with the DDAH to investigate and protect the archaeological site of Faida through the definition of a fenced core zone and a buffer zone, to conserve its reliefs, and make them safe for outdoor display. In October 2022, the site was opened to the public and has since been visited by thousands of tourists coming from Kurdistan, Iraq, and abroad.

In 2023, thanks to the relentless support of AICS, the executive stage of the project was launched in strict cooperation with the DDAH to create 'Sennacherib's Archaeological Park'. At the heart of the project is the firm belief that the heritage sites of the Northern Assyrian Irrigation System are critical assets for developing a sustainable Kurdish and Iraqi tourism sectors. Such assets will enhance employment opportunities for young people living in rural areas and the city of Duhok as specialised workmen for archaeological excavations and for building and maintaining the archaeological park. The second ongoing stage of the project includes the full conservation of the Maltai, Faida, and Khinis rock reliefs and the Jerwan aqueduct, the creation of itineraries for tourists, the implementation of lighting and security systems, the establishment of reception and surveillance units, the building of parking areas, and the creation of on-site durable multilingual information panels and multimedia contents accessible through QR code (in Kurdish, Arabic, and English).

Capacity-building workshops for tourist guides for the archaeological park, tourism promotion, and marketing are delivered to groups of unemployed young people including members of the different communities living in the region (Kurds, Arabs, Christians, Yazidis). Local workmen residing in the villages in the environs of the archaeological sites are being trained on-the-job at Khinis, Jerwan, Maltai, and Faida with the aim of providing them with new skills to become specialised in the maintenance of archaeological sites and parks and in the archaeological excavation techniques. Finally, employees of the DDAH are trained on-the-job by LoNAP to monitor the state of sculpted rock surfaces, in basic rock relief conservation techniques, the field documentation of archaeological park monuments using UAV and laser scanning technologies, archaeological excavation techniques, and the management and enhancement of cultural heritage.

6.5 Community Engagement and Empowerment: Towards a More Inclusive and Multivocal Archaeological Park

Heritage is a valuable resource, not only economically (through jobs, earnings, and business development) but also in terms of individual and societal well-being, encompassing education, identity, civic cohesion, and a sense of belonging (Taçon and Baker, 2019; Burtenshaw, 2014). However, local communities often remain detached from their heritage, missing out on opportunities linked to sustainable development. The direct involvement of host communities is now recognised as essential in sustainable heritage management (Jameson, 2019), with recent studies demonstrating the effectiveness of community engagement in sustainable heritage tourism (Moscardo, 2008; Hall and Richards, 2000; Rasoolimanesh and Jaafar, 2017; Su and Wall, 2014).

International institutions like UNESCO increasingly recognise the crucial role of local communities in heritage conservation and management. This shift is evident in initiatives such as the Budapest Declaration (2002), where the World Heritage Committee introduced the 'four Cs' – conservation, credibility, communication, and capacity building[6] – adding a fifth objective, communities, in 2007 to emphasise the involvement of Indigenous peoples and local communities (UNESCO, 2024d). The Faro Convention (2005) also places a strong emphasis on people and human values in the context of cultural heritage, highlighting its potential for sustainable development. The 40th anniversary of the World Heritage Convention in 2012, themed 'Local Community and Sustainable Development', further reinforced this focus, culminating in the Kyoto Vision declaration, which called for substantial financial resources and capacity building for global heritage conservation (UNESCO, 2024e). The renewed focus by international institutions on the crucial role of local communities aligns with significant developments in academic research and professional practice (Gould, 2019, pp. 20–26). Researchers are now systematically exploring the approaches, methods, and outcomes of projects related to heritage protection, community engagement, and sustainable development (e.g. Gould, 2019; Pyburn, 2020; Higgins and Douglas, 2020; Baillie and Sørensen, 2021; Labadi *et al.*, 2020; Labadi, 2022). However, findings often reveal that many projects struggle with sustainability, highlighting that a 'one-size-fits-all' formula is ineffective and that each project must adapt to the political, economic, and social context that hosts the heritage.

The Sennacherib's Archaeological Park Project has primarily followed a top-down approach because the project was designed and proposed by a foreign university (the University of Udine) and the main partner in its implementation is a government institution (the Directorate of Antiquities and Heritage of Duhok). This approach is largely determined by the funders who support and have supported the project: AICS, the Italian Ministry of Foreign Affairs and International Cooperation, the ALIPH Foundation, and the Gerda Henkel Stiftung. At the same time, the poor preservation and the threats endangering the unique archaeological sites that the project intends to safeguard and restore required prompt interventions that are not achievable via a more desirable – but at the same time much slower – bottom-up approach. However, since its inception, the archaeologists, architects, restorers, and the rest of the team have been trying to mitigate this.

Since its beginning, LoNAP has always been active in what is defined as public archaeology: the 'practice and scholarship of where archaeology meets the world' (Moshenska, 2017, pp. 1–13). This has involved engaging with the local and international public to spread information about the project's activities and results, utilising social media posts, press conferences, meetings with local and national authorities, and TV documentaries.

Starting in 2021, LoNAP embraced a broader view of community archaeology: 'archaeology by the people for the people' (Reid, 2012, p. 18). This has the specific aim of engaging with all communities living within the boundaries of Sennacherib's Archaeological Park and all the stakeholders involved (Simi et al., in press). The Community Archaeology Programme was designed and implemented with the project's main partner, the DDAH.

One of the first steps was to raise awareness of the significance and fragility of the local archaeological sites and the archaeological excavations, documentation, and restoration that LoNAP and DDAH were conducting. For this reason, we organised a public meeting at the Cultural Centre of the town of Faida, several open days and site visits with the archaeologists accompanied by numerous activities for the primary and secondary schools at Faida and Khinis. To encourage awareness building and our inclusion efforts, we prepared a multilingual virtual visit to the archaeological site of Faida[7] (that can also be used as an audio guide to the site) and a series of information brochures for children and adults (in Kurdish, Arabic, and English). Open days with the team and activities for children and college students were often organised at Khinis. The site visits and the community meetings were not only important opportunities to create relationships with the people participating in the activities, but also opportunities to collect the perceptions, memories, and interpretations of the local communities either through questionnaires (at Faida in a pilot season in 2022) or via unstructured informal interviews.

Systematic ethnographic research[8] is currently taking place in the villages surrounding the archaeological sites of Khinis (Khinis and Bavian), Maltai (Qassara), and Jerwan (Mahad). The goal is to investigate the relationships between the local people and these sites, including their active engagement with the heritage, whether they have a sense of ownership and pride in the heritage, the way they use it and engage with it. Additionally, the research aims to explore the community's aspirations for the future of the sites, such as potential touristic paths, job and business opportunities, and their overall use. This ethnographic research, along with the numerous activities at the sites, will provide the local community with the opportunity to give voice to their opinions, ask questions, and be involved in the decision-making process regarding future plans for the enhancement of the archaeological sites.

To support sustainability beyond the end of the project, several different stakeholders are involved in the community engagement activities (Figure 6.4). Important stakeholders are the three universities of Duhok (Duhok University, Newroz University, and American University of Duhok). To engage them in the archaeological park's design and implementation, in October 2023, we organised a series of field activities at Khinis and Jerwan and a Masterclass reserved for the students of Architecture and Planning on the topics of Archaeological Park Design and Creation.

Since October 2023, we are also working on a Memorandum of Understanding between the University of Duhok[9] and the Sennacherib's Archaeological Park Project to establish a cooperation programme that includes the involvement of several departments in the implementation of the park (Departments of Architecture, Planning, Archaeology, Conservation and Eco-Tourism). The agreement will enable the organisation of shared activities on sites and the participation of students as interns in the project.

The other important stakeholders, crucial for the future of the archaeological park, are the KRG General Directorate of Antiquities and Heritage, the Iraqi State Board of Antiquities, the KRG Ministry of Municipality and Tourism, and the Governorate of Duhok. These stakeholders are constantly involved in the archaeological park's progress and are often engaged in events at the sites such as the 2022 'Opening of the Faida Archaeological Park' and the 2023 'Revealing of the Restored Great Relief at Khinis'. These events, open to the public and authorities, were important moments to connect the different communities living in the area with the archaeologists, architects, and conservators working on the creation of the archaeological park, local and foreign authorities, and the media.

Our goal is to establish a stronger, more inclusive, and even more participative community engagement programme in the sites involved in the Sennacherib's Archaeological Park Project. The final aim is to challenge – or at least mitigate – the power imbalance between the archaeologists and heritage practitioners and the local communities, leading to closer collaboration that will lay the foundation

Figure 6.4 Photomosaic of community archaeology activities carried out by LoNAP and DDAH. (a) College students from Duhok visiting Faida in October 2021, (b) students from the Dohuk Fine Arts Institute preparing their paintings for the First Khinis Art Award in October 2023, (c) high school students from Domiz visiting Faida after the opening of the archaeological park in October 2022, and (d) children from the Khinis elementary school playing and learning about the Khinis Archaeological Complex in October 2023. All images © LoNAP.

for the residents to take charge of their own development trajectories. To achieve this, it is necessary to co-create a vision for cultural tourism in the Governorate of Duhok aimed at linking and motivating the local stakeholders and fostering their collaboration with each other.

6.6 Reflections

One of the central debates in community archaeology since its emergence is identifying the relevant 'community' or 'communities' involved (e.g. Atalay, 2012, pp. 90–95; Marshall, 2002; Moshenska and Dhanjal, 2012; Smith and Waterton, 2009). In the KRI, this task is particularly challenging. Although the country is predominantly Sunni Kurdish, it has significant religious and ethno-religious diversity. Since 2003, the KRI has seen an influx of minority groups fleeing instability and violence from Iraq's south and west, especially during Daesh's campaigns between 2014 and 2017. This has made the KRI a refuge for minority populations in the region. Today, the KRI is home to most of Iraq's Christian populations, including Assyrians, Chaldeans, and Syriacs, as well as almost the entire Yazidi community. It also houses Shi'a and Sunni Arabs, Turkmen, Shabak, Kaka'i, Zoroastrians, and small numbers of Sabean Mandaeans and Baha'i (Smith and Shadarevian, 2017). Iraq's political crises have complicated the peaceful coexistence of these diverse communities, exacerbating sectarianism and the exploitation of cultural heritage by local political elites (Kathem *et al.*, 2022, p. 32).

Heritage can be the cause of social tension and conflict (Bsheer, 2017). The main factors creating tensions around archaeological sites are preservation and knowledge creation, economic development, and social use for local communities (Gilliot, 2020, p. 28). Conflict can arise from the intrinsic dissonance of the heritage, when multiple groups attribute different values to or claim ownership of the same heritage (Gilliot, 2020; Graham et al., 2000; Graham et al., 2005; Tunbridge and Ashworth, 1996). Assyrian heritage represents a dissonant – and to some extent disputed – heritage for the modern Assyrian and Kurdish communities (Edward, 2020).

However, cultural heritage and archaeology are also powerful tools for promoting peacebuilding, reconciliation, and social cohesion in post-conflict scenarios (Alsalloum and Brown, 2019; Giblin, 2014; Lostal and Cunliffe, 2016; Matthews *et al.*, 2020; Newson and Young, 2017). The creation of Sennacherib's Archaeological Park aims to celebrate diversity and encourage dialogue and mutual understanding among different cultural, ethnic, and religious groups.

As mentioned earlier, a further challenge – but also a future opportunity – is inscribing Sennacherib's irrigation network and its monuments in the UNESCO World Heritage List. Despite the criticisms of the World Heritage concept (Labadi, 2013; Meskell, 2018) and the unjust nomination system of UNESCO,[10] the DDAH and many local stakeholders are interested in and committed to putting maximum effort into submitting a World Heritage List application. Unfortunately, the present state of strained political relations between the KRI and the Federal Republic of Iraq is making it difficult to initiate the submission process, which should be led by the central authorities in Baghdad. Studies suggest that while listing as a WHS will increase the protection and conservation of Sennacherib's irrigation network, it will also increase the number of visitors and economic development of the host region (e.g. Hosseini *et al.*, 2021; Yang *et al.*, 2019; Su and Lin, 2014). This will require enhanced efforts to maintain and sustain the heritage and natural landscape.

Climate change provides a further challenge that Sennacherib's Archaeological Park and the KRI will soon face, a formidable source of stress for West Asian countries (Seyfi and Hall, 2021, pp. 322–324). In the KRI, the ongoing widespread semi-aridification process (Khayyat et al., 2020) coupled with rapid population growth and environmental degradation strongly increases the vulnerability of the region to further climate change (IOM, 2024, p. 4–6). This situation will likely increase the tension between tourism and other sectors over water resources, as tourism is a major user of water, both directly and indirectly (Hall, 2019). Although the KRG has a clear interest in expanding tourism in its region, as yet there has been no attempt to promote tourism-related adaptation and mitigation.

In the context of the generalised reduction in water supply affecting the KRI, the Assyrian canal network in Nineveh's hinterland that tapped water from rivers, wadis, and karst springs, represents a highly sustainable irrigation infrastructure that should be taken as a model for renewal of the present crop irrigation system based on the indiscriminate excavation of boreholes/wells that severely deplete water tables and the uncontrolled extraction of water from rivers/streams (IOM, 2024, p. 9). Rising temperatures, changing rainfall patterns, accelerating evapotranspiration, and increased frequency and intensity of drought are dramatically impacting water sources for agriculture and human consumption.[11] In such a severely degraded and potentially explosive context, traditional water exploitation systems and techniques, which developed over a long period by trial and error and adaptation to ever-changing circumstances, represent a lesson to (re)learn in terms of functionality and sustainability. The protection and enhancement of the Northern Assyrian Irrigation System through the creation of Sennacherib's Archaeological Park and the development of sustainable tourism express these strategic values, while also contributing to raising awareness about the need for new agricultural policies and practices in Iraq and beyond.

Acknowledgements

We are grateful to the General Directorate of Antiquities of the KRG, the DDAH and their teams for their support to our project. We also are indebted to the editors of this book, Steven Mithen, Maria Rabbani, and Mubariz Ahmed Rabbani, for editing our English text and useful suggestions that have improved our article. The LoNAP is supported by: University of Udine, Italian Ministry of Foreign Affairs and International Cooperation, Friuli Venezia Giulia Regional Authority, Friuli Banking Foundation, Italian Ministry of Universities and Research, and Italian Agency for Development Cooperation. This paper was written in the framework of the PRIN 2022 Project 'Near Eastern Empires at Work (NEEW), an archaeological approach to the study of empires, their organisation and impact on local societies and landscapes in northern Iraq' (CUP: G53D23000300006) funded by the European Union – NextGenerationEU Initiative.

Notes

1 All authors contributed to the research conception and design. All authors reviewed, commented, and agreed on the entire manuscript. Sections 1 and 6 were written by D.M.B, F.S., and B.J.H.; sections 2 and 5 are the work of F.S.; and sections 3 and 4 were drafted by D.M.B.

2 The document is available at: www.iraq-businessnews.com/wp-content/uploads/2023/09/KRG-Reform-Roadmap-for-Tourism-Sector-of-KRI-2022.pdf (Accessed on 08/06/24).

3 Kurdistan Regional Parliament – Iraq (2021), 'Parliament passes Law on Management and Protection of Archaeology and Heritage in Kurdistan Region', 30 April 2021, www.parliament.krd/english/parliament-activities/latest-news/posts/2021/april/parliament-passes-law-on-management-and-protection-of-archeology-andheritage-in-kurdistan-region.

4 The Central Zagros Archaeological project is directed by Professor Roger Matthews and Dr Wendy Matthews at the University of Reading. Full details are available at its website: www.czap.org/HESDIK (Accessed on 08/06/24).

5 The "Land of Nineveh Archaeological Project (LoNAP)" is a wide-range multidisciplinary research project carried out by the Italian Archaeological Mission in the Kurdistan Region of Iraq (IAMKRI). of the University of Udine. Full details can be found at its website: www.terradininive.com/land-of-niniveh/?lang=en (Accessed on 08/06/24).

6 See the Budapest Declaration on World Heritage (WHC-02/CONF.202/5) – http://whc.unesco.org/document/1334.

7 The virtual tour was implemented by the University of Udine in cooperation with A. Cecili and D. Coppola of the Digital Earth Lab of the University of Roma Tre and thanks to the technical support of 3D Target srl. The virtual tour can be accessed at: www.terradininive.com/visita-virtuale-di-faida.

8 The ethnographic research is conducted by Y. Nabati Mazloumi of Glasgow University.

9 The 'Sennacherib's Archaeological Park' Project Coordinator is an architect from Duhok University, Saman Kareem, who is also actively involved in the design and implementation of the park.

10 A simple comparison of the numbers of WHSs located in different parts of the world reveals an unfair distribution. As of 2024, over 47% of WHSs are located in Europe and North America, while less than 8% are in the Arab States, and less than 9% are in Africa. Regarding the list of WHSs in danger, only 13% are in Europe and North America, while the majority (e.g. 41% in the Arab States and 25% in Africa) are in the Global South. Data from UNESCO 2024c.

11 A report by the Norwegian Refugee Council (NRC) shows that extreme weather affecting all of Iraq is negatively impacting crop yields, and market system access and function, exacerbating social schisms and precipitating risks of secondary displacement (NRC, 2023).

References

Alsalloum, A. and Brown, A. (2019). Towards a Heritage-Led Sustainable Post-Conflict Reconciliation: A Policy-Led Perspective, *Sustainability,* 11(6), pp. 1686.

Atalay, S. (2012). *Community-Based Archaeology: Research with, by, and for Indigenous and Local Communities*. Berkeley: University of California Press.

Baillie, B. and Sørensen, M.L.S. (2021). *African Heritage Challenges: Communities and Sustainable Development*. Singapore: Springer.

Bsheer, R. (2017). Heritage as War, *International Journal of Middle East Studies,* 49, pp. 729–734.

Burtenshaw, P. (2014). Mind the Gap: Cultural and Economic Values in Archaeology, *Public Archaeology,* 13 (1–3), pp. 48–58.

Burtenshaw, P. and Palmer, C. (2014). Archaeology, Local Development and Tourism—a Role for International Institutes, *Bulletin of the Council for British Research in the Levant*, 9(1), pp.21–26.

Cernea M. (2001). *Cultural Heritage and Development: A Framework for Action in the Middle East and North Africa*. Washington: World Bank Group.

Comer, D. C. and Willems W.J.H. (2011). *Tourism and Archaeological Heritage: Driver to Development or Destruction?* Available at: https://openarchive.icomos.org/id/eprint/1208/ (Accessed: 8 June 2024).

Council of Europe (2005). *Convention on the Value of Cultural Heritage for Society (Faro Convention, 2005)*. Available at: https://rm.coe.int/1680083746 (Accessed: 8 June 2024).

Deacon, J. (2006). Rock art Conservation and Tourism, *Journal of Archaeological Method and Theory,* 13 (4), pp. 376–396.

Düring, B.S. (2020). *The Imperialisation of Assyria: An Archaeological Approach*. Cambridge: University Press.

Edward, R. (2020). *Assyrian Cultural Heritage at Risk in Northern Iraq*. Australia: Assyrian Cultural & Social Youth Association Inc. Available from: https://static1.squarespace.com/static/6455aea26b1d6271277e1c0c/t/649e37338aa15f0763b7ddc5/1688090431918/Assyrian+Cultural+Heritage+at+Risk+in+Northern+Iraq.pdf (Accessed: 8 June 2024).

Fales, F.M. and Del Fabbro, R. (2014). Back to Sennacherib's Aqueduct at Jerwan: A Reassessment of the Textual Evidence, *Iraq,* 76, pp. 65–98.

Giblin, J.D. (2014). Post-Conflict heritage: Symbolic Healing and Cultural Renewal, *International Journal of Heritage Studies*, 20(5), pp. 500–518.

Giliberto, F. and Labadi, S. (2022). Harnessing Cultural Heritage for Sustainable Development: An Analysis of Three Internationally Funded Projects in MENA Countries, *International Journal of Heritage Studies*, 28(2), pp. 133–146.

Gilliot L. (2020). 'Archaeologists and Tourism: Symbiosis or Contestation?', in: D. J. Timothy and L. G. Tahan (eds.) *Archaeology and Tourism, Touring the Past, 2020*. Bristol: Channel View Publications, pp. 26–40.

Gould, P.G. (2019) *Empowering Communities through Archaeology and Heritage. The Role of Local Governance in Economic Development*. New York: Bloomsbury Publishing.

Gould, P.G. and Burtenshaw, P. (2019). 'Community Economic Development and Archaeology', in: *Encyclopedia of Global Archaeology*. DOI: https://doi.org/10.1007/978-3-319-51726-1_3385-1 (Accessed: 8 June 2024).

Graham, B., Ashworth, G.J., and Tunbridge, J.E. (2000). *A Geography of Heritage: Power, Culture and Economy*. London: Arnold.

Graham, B., Ashworth, G.J., and Tunbridge, J.E. (2005). 'The Uses and Abuses of Heritage', in: G. Corsane (ed.) *Heritage, Museums and Galleries: An Introductory Reader*. London: Routledge, pp.28–40.

Grayson, A.K. and Novotny, J. (2014). *The Royal Inscriptions of Sennacherib, King of Assyria (704-681 BC), Part 2*. Indiana: Eisenbrauns.

Hall, C.M. (2019). '*Tourism and Climate Change in the Middle East*', in: D. Timothy (ed.) *The Routledge handbook of tourism in the Middle East and North Africa.* Abingdon: Routledge, pp. 109–209.

Hall, D. and Richards, G. (2000). *Tourism and Sustainable Community Development*. London: Routledge.

Hampton, M.P. (2005). Heritage, Local Communities and Economic Development, *Annals of Tourism Research*, 32(3), pp. 735–759.

Higgins, V. and Douglas, D. (2020). *Communities and Cultural Heritage: Global Issues, Local Values*. London: Routledge/Taylor & Francis Group.

Hosseini, K., Stefaniec, A., and Hosseini, S.P. (2021). World Heritage Sites in Developing Countries: Assessing Impacts and Handling Complexities Toward Sustainable Tourism, *Journal of Destination Marketing & Management*, 20, pp. 100–616.

IOM – International Organization for Migration (2024). *Climate Vulnerability Assessment in the Kurdistan Region of Iraq*. Available at: https://reliefweb.int/report/iraq/climate-vulnerability-assessment-kurdistan-region-iraq-data-collection-period-september-december-2023 [Accessed: 8 June 2024].

Jacobsen, T. and Lloyd, S. (1935). *Sennacherib's Aqueduct at Jerwan*. Chicago: University of Chicago Press.

Jameson, J.H. (2019). 'Introduction: The Critical Junctures of Archaeology, Heritage, and Communities', in: J. H. Jameson and S. Musteaţă (eds.) *Transforming Heritage Practice in the 21st Century: Contributions from Community Archaeology*. Cham: Springer International Publishing, pp. 1–12.

Jarah, S.H.A., Zhou, B., Abdullah, R.J., Lu, Y., and Yu, W. (2019). Urbanization and Urban Sprawl Issues in City Structure: A Case of the Sulaymaniah Iraqi Kurdistan Region, *Sustainability*, 11(2), pp. 1–21.

Kathem, M., Robson, E., and Tahan, L.G. (2022). *Cultural Heritage Predation in Iraq. The Sectarian Appropriation of Iraq's Past*. Research papers [Chatham House]. DOI: https://doi.org/10.55317/9781784135119 (Accessed: 8 June 2024).

Khayyat, H.A.K.A., Sharif, A.J.M., and Crespi, M. (2020). 'Assessing the Impacts of Climate Change on Natural Resources in Erbil Area, the Iraqi Kurdistan Using Geo-Information and Landsat Data', in: A. Al-Quraishi, and A.Negm (eds.) *Environmental Remote Sensing and GIS in Iraq*. Cham: Springer, pp. 463–498.

Kopanias, K. and MacGinnis, J. (2016). *The Archaeology of the Kurdistan Region of Iraq and Adjacent Regions*. Oxford: Archaeopress.

Kopanias, K., MacGinnis, J. and Ur, J.A. (2015). *Archaeological Projects in the Kurdistan Region in Iraq*. Erbil: General Directorate of Antiquities in the Kurdistan Region in Iraq.

KRSO – Kurdistan Region Statistics Office (2024). Population. Available at: https://krso.gov.krd/en/indicator/population-and-labor-force/population [Accessed: 8 June 2024]

Labadi, S. (2013). *UNESCO, Cultural Heritage, and Outstanding Universal Value: Value-Based Analyses of the World Heritage and Intangible Cultural Heritage Conventions*. Lanham: Rowman & Littlefield Pub.

Labadi, S. (2019). *The Cultural Turn in International Aid: Impacts and Challenges for Heritage and the Creative Industries*. London: Routledge.

Labadi, S. (2022). *Rethinking Heritage for Sustainable Development. International Framework, Local Impacts*. London: UCL Press.

Labadi, S., Giliberto, F., Rosetti, I., Shebati, L., and Yildirim, E. (2021). *Heritage and the Sustainable Development Goals. Policy Guidance for Heritage and Development Actors*. Paris: ICOMOS.

Labadi, S. Giliberto, F. Taruvinga, P., and Jopela, A. (2020). *World Heritage and Sustainable Development in Africa: Implementing the 2015 Policy*. Midrand: African World Heritage Fund.

Lenik, S. (2013). Community Engagement and Heritage Tourism at Geneva Estate, Dominica, *Journal of Heritage Tourism,* 8(1), pp. 9–19.

Lostal, M. and Cunliffe, E. (2016). Cultural Heritage that Heals: Factoring in Cultural Heritage Discourses in the Syrian Peacebuilding Process. *Historic Environment, Policy and Practice,* 7(2–3), pp. 248–259.

Mardas, J. (2017). Iraqi Kurdistan's Heritage in the Face of Regional Development: State of Preservation of Archaeological Sites and Damage Assessment – Preliminary Report' *Polish Archaeology in the Mediterranean,* 26(1), pp. 809–826.

Marshall, Y. (2002). What Is Community Archaeology?, *World Archaeology,* (34)2, pp. 211–219.

Matsuda, A. and Okamura, K. (2011). 'Introduction: New Perspective in Global Public Archaeology,' in: K. Okamura and A. Matsuda (eds.) *New Perspectives in Global Public Archaeology*. London: Springer, pp. 1–18.

Matthews, R., Rasheed, Q. H., Palmero Fernandez, M., Fobbe, S., Novacek, K., Mohammed-Amin, R., Muhl, S., and Richardson, A. (2020). Heritage and Cultural Healing: Iraq in a Post-Daesh Era', *International Journal of Heritage Studies,* 26 (2), pp. 120–141.

Meskell, L. (2018). *A Future in Ruins: UNESCO, World Heritage, and the Dream of Peace.* New York, NY: Oxford University Press.

Morandi Bonacossi, D. (2018a). 'The Creation of the Assyrian Heartland: New Data from the 'Land Behind Nineveh'', in: B.S. Düring and T. Stek (eds.) *The Archaeology of Imperial Landscapes. A Comparative Study of Empires in the Ancient Near East and Mediterranean World.* Cambridge: Cambridge University Press, pp. 48–85.

Morandi Bonacossi, D. (2018b). 'Water for Nineveh. The Nineveh Irrigation System in the Regional Context of the 'Assyrian Triangle': A First Geoarchaeological Assessment', in: H. Kühne (ed.) *Water for Assyria*, Studia Chaburensia 7. Wiesbaden: Harrassowitz, pp. 77–115.

Morandi Bonacossi, D. (2018c). Twelve Royal Stelas for Twelve Great Gods: New Discoveries at the Khinis Monumental Complex, *Ash-Sharq,* 2(2), pp. 76–97.

Morandi Bonacossi, D. (2021). 'Faces in the Stone. Rock Art in the Duhok Region (Kurdistan Region of Iraq). A Seriously Threatened Cultural Heritage', in: A. Traviglia, L. Milano, C. Tonghini, R. Giovanelli (eds.) *Stolen Heritage Multidisciplinary Perspectives on Illicit Trafficking of Cultural Heritage in the EU and the MENA Region.* Venice: Venice University Press, pp. 35–78.

Morandi Bonacossi, D. and Iamoni, M. (2015). Landscape and Settlement in the Eastern Upper Iraqi Tigris and Navkur Plains: The Land of Nineveh Archaeological Project, Seasons 2012–2013, *Iraq*, 77, pp. 9–39.

Morandi Bonacossi, D. and Qasim, H.A. (2022). Irrigation and Landscape Commemoration in Northern Assyria. The Assyrian Canal and Rock Reliefs in Faida (Kurdistan Region of Iraq). Preliminary Report on the 2019 Field Season, *Iraq*, 84, pp. 43–81.

Morandi Bonacossi, D. and Tonghini, C. (2018). The Many Facets of Heritage Destruction in Iraq, *KASKAL*, 15, pp. 267–76.

Moscardo, G., (2008). *Building Community Capacity for Tourism Development.* UK: CABI.

Moshenska, G. (2017). 'Introduction: Public Archaeology as Practice and Scholarship Where Archaeology Meets the World', in: G. Moshenska (ed.) *Key Concepts in Public Archaeology.* London: UCL Press, pp. 1–13.

Moshenska, G. and Dhanjal, S. (2012). 'Introduction: Thinking About, Talking About, and Doing Community Archaeology', in: G. Moshenska and S. Dhanjal (eds.) *Community Archaeology: Themes, Methods and Practices.* Oxford: Oxbow Books, pp. 1–5.

Newell, G.E. (2008). Rhyming Culture, Heritage, and Identity: The 'Total Site' of Teotihuacan, Mexico, *Journal of Heritage Tourism,* 3 (4), pp. 243–255.

Newson, P. and Young, R. (2017). *Post-Conflict Archaeology and Cultural Heritage: Rebuilding Knowledge, Memory and Community from War-Damaged Material Culture.* London: Routledge.

NRC – Norwegian Refugee Council (2023). *Inadequate and Inequitable: Water Scarcity and Displacement in Iraq.* Norwegian Ministry of Foreign Affairs.

Orazi, R. (2019). *The Archaeological Environmental Park of Sennacherib's Irrigation Network. Recording, Conservation and Management of the Cultural Heritage of the Northern Region of Iraqi Kurdistan.* Udine: Forum Editrice.

Osman, M. (2021). *Kurdistan Region of Iraq: Population Analysis Report.* KRSO (Erbil). Available at: https://krso.gov.krd/content/upload/1/root/kurdistan-population-analysis-report-6-english-final-corrected-29052022.pdf (Accessed: 8 June 2024).

Pyburn, K.A. (2020). 'Archaeology, Communities and Economic Development: A Pandora's Box', in: V. Higgins and D. Douglas (eds.) *Communities and Cultural Heritage: Global Issues, Local Values*. London: Routledge/Taylor & Francis Group, pp. 93–105.

Rasoolimanesh, S.M. and Jaafar, M. (2017). 'Sustainable Tourism Development and Residents' Perceptions in World Heritage Site Destinations', *Asia Pacific Journal of Tourism Research*, 22 (1), pp. 34–48.

Reid, P. (2012). 'Performance or Participation: The Relationship Between Local Communities and the Archaeological Domain', in: G. Moshenska and S. Dhanjal (eds.) *Community Archaeology: Themes, Methods and Practices*. Oxford: Oxbow Books, pp. 18–27.

Sabr, C.A. (2014). A Study on the Urban Form of Erbil City (The Capital of Kurdistan Region) as an Example of Historical and Fast Growing City, *Humanities and Social Sciences Review*, 3, pp. 325–40.

Seyfi, S. and Hall, C.M. (2021). *Cultural and Heritage Tourism in the Middle East and North Africa: Complexities, Management and Practices*. Abingdon/New York: Routledge.

Simi, F., Yaseen, K., and Morandi Bonacossi, D. (in press). Community Archaeology Challenges and Perspectives from the Kurdistan Region of Iraq: The Case of Faida (Duhok Governorate), *ARCHEOSTORIE® Journal of Public Archaeology*.

Sinha, A., Kathayat, G., Weiss, H., Li, H., Cheng, H., Reuter, J., Schneider, A.W., Berkelhammer, M. Adalı, S.M., Stott, L.D., and Edward, R.L. (2019). Role of Climate in the Rise and Fall of the Neo-Assyrian Empire Factors in the Decline of the Neo-Assyrian Empire, *Science Advances,* 5, pp. 1–10.

Smith, L. and Waterton, E. (2009). *Heritage, Communities and Archaeology*. London: Duckworth Academia.

Smith, M.I.C and Shadarevian, V. (2017). *Wilting in the Kurdish Sun: the Hopes and Fears of Religious Minorities in Northern Iraq*. Report for the United States Commission on International Religious Freedom. Available at: www.uscirf.gov/sites/default/files/Kurdistan%20Report%20for%20printing.pdf (Accessed: 8 June 2024).

Stansfield, G.R.V. (2003). *Iraqi Kurdistan: Political Development and Emergent Democracy*. London: Routledge.

Su, M.M., and Wall, G. (2014). Community Participation in Tourism at a World Heritage Site: Mutianyu Great Wall, Beijing, China, *International Journal of Tourism Research*, 16(2), pp. 146–156.

Su, Y.W. and Lin, H.L. (2014). Analysis of International Tourist Arrivals Worldwide: The Role of World Heritage Sites, *Tourism Management*, 40(C), pp. 46–58.

Taçon, P.S.C. and Baker, S. (2019). New and Emerging Challenges to Heritage and Well-Being: A Critical Review, *Heritage*, 2(2), pp.1300–1315.

Timothy, D.J. (2014). Contemporary Cultural Heritage and Tourism: Development Issues and Emerging Trends, *Public Archaeology*, 13(1–3), pp. 30–47.

Timothy, D.J. (2019). '*Understanding the Middle East and North Africa*', in: D.J. Timothy (ed.) *Routledge Handbook on Tourism in the Middle East and North Africa*. Abingdon: Routledge, pp 3–10.

Timothy, D.J and Nyaupane, G.P. (2009). *Cultural Heritage and Tourism in the Developing World: A Regional Perspective*. London: Routledge.

Timothy, D.J. and Tahan, L. (2020). 'Archaeology and Tourism: Consuming, Managing and Protecting the Human Past', in: D.J. Timothy and L.G. Tahan (eds.) *Archaeology and Tourism, Touring the Past*. Bristol: Channel View Publications, pp. 1–25.

Tunbridge, J.E. and Ashworth, G.J. (1996). *Dissonant Heritage: The Management of the Past as a Resource in Conflict*. Chichester: Wiley.

UNESCO (2024a). *World Heritage List*. Available at: https://whc.unesco.org/en/list/ (Accessed: 8 June 2024).

UNESCO (2024b). *World Heritage Tentative Lists*. Available at: https://whc.unesco.org/en/tentativelists/ (Accessed: 8 June 2024).

UNESCO (2024c). *World Heritage List Statistics*. Available at: https://whc.unesco.org/en/list/stat#s9 (Accessed: 8 June 2024).

UNESCO (2024d). *The World Heritage Convention*. Available at: https://whc.unesco.org/en/convention/ (Accessed: 8 June 2024).

UNESCO (2024e). *The Kyoto Vision*. Available from: https://whc.unesco.org/document/123339 (Accessed: 8 June 2024).

Ur, J. (2005). Sennacherib's Northern Assyrian Canals. New Insights from Satellite Imagery and Aerial Photography, *Iraq,* 67, pp. 317–345.

Ur, J. (2017). The Archaeological Renaissance in the Kurdistan Region of Iraq, *Near Eastern Archaeology*, 80(3), pp. 176–187.

Ur, J., Babakr, N., Palermo, R., Creamer, P., Soroush, M., Ramand, Sh., and Nováček, K. (2021). The Erbil Plain Archaeological Survey: Preliminary Results, 2012-2020, *Iraq,* 83, pp. 205–243.

Ur, J.A., de Jong, L., Giraud, J., Osborne, J.F., and MacGinnis, J. (2013). Ancient Cities and Landscapes in the Kurdistan Region of Iraq: The Erbil Plain Archaeological Survey 2012 Season, *Iraq* 75, pp. 89–118.

Ur, J.A. and Reade, J.E. (2015). The Hydraulic Landscape of Nimrud, *Mesopotamia,* 50, pp. 25–51.

Wall, G. and Mathieson, A. (2006). *Tourism: Changes, Impacts and Opportunities*. Harlow: Prentice Hall.

Wilkinson, T. J. (2004). *On the Margin of the Euphrates: Settlement and Land Use at Tell es-Sweyhat and in the Upper Lake Assad Area, Syria.* Chicago: Oriental Institute of the University of Chicago.

Yang, Y., Xue, L., and Jones, T.E. (2019). Tourism-enhancing effect of World Heritage Sites: Panacea or placebo? A meta-analysis, *Annals of Tourism Research*, 75, pp. 29–41.

7 Building Community Archaeology Practice and Heritage Tourism for Sustainable Development in a Post-Conflict Zone

The ACT and ALIPH Projects in the Swat Valley (Pakistan)

Luca M. Olivieri

7.1 The Swat Valley of Pakistan

The Swat Valley in northern Pakistan forms part of the Malakand Division of the Khyber Pakhtunkhwa Province (formerly the North-West Frontier Province of Pakistan). The valley lies at an average altitude of 1,000 metres asl in a mountainous region of the Karakoram–Hindukush piedmont. It is formed by the course of the Swat River which originates from the glaciers of the Upper Swat mountains (Swat Kohistan) and is characterised by extensive fertile alluvial soils formed from Pleistocene and Early Holocene deposits rich in nutrients and minerals (Figure 7.1). Unlike other valleys in the region, Swat is oriented mostly E-W, giving it superior sun exposure. Because this exposure is combined with fertile soils, a perennial water supply, and protection from both monsoons and winter winds from the north, Swat has a high level of agricultural production based on double-cropping: wheat and barley in the spring and rice, millet, and maize in the autumn.

Throughout the Holocene, this productivity combined with its forest and mineral resources has provided Swat with a stable and important resource base, one of the most productive in the north-west of the Asian subcontinent. This has been exploited by fortified agricultural settlements since the Bronze Age, by the Achaemenid Dynasty (559–330 BCE), throughout the pre-modern era, and still today, with the major cultural phases of Swat's history linked to climatic optima. A consequence is that the Swat Valley has one of the most important archaeological regions in South Asia, as reviewed by Olivieri (2022).

When Swat was part of the Achaemenid satrapy (province) of Gandara (Gandhara), the valley was briefly occupied by Alexander the Great. Between the 3rd century BCE and the 7th–8th century CE, the ancient cultures of Swat had a considerable impact on the history of Buddhism. It was in Swat that the Buddhist narrative art known as Gandharan art was developed (1st–3rd century CE) and where the Late Buddhist Vajrayana schools were established (7th–8th century CE).

DOI: 10.4324/9781003491071-8

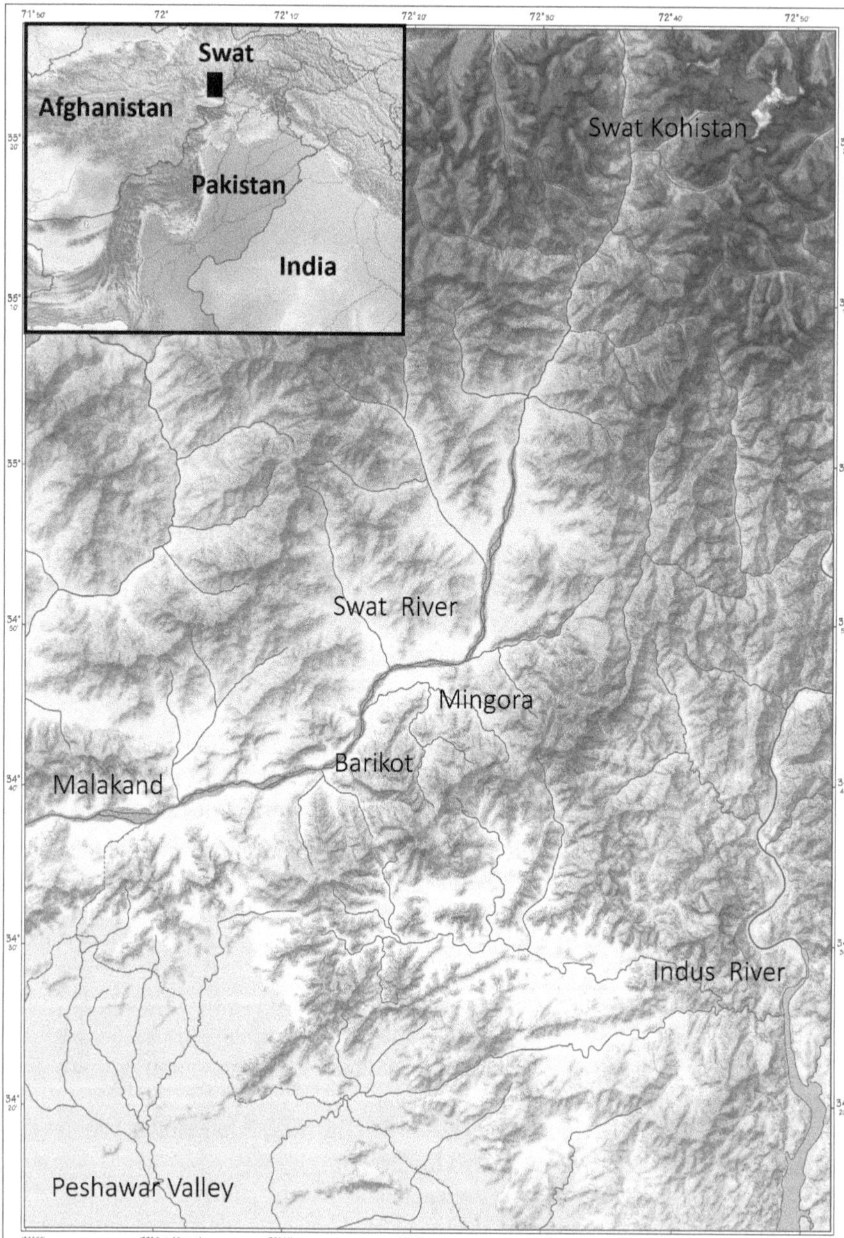

Figure 7.1 Swat Valley in northern Pakistan, showing the location of Barikot. © ISMEO and Department of Geography, University of Vienna.

These schools were important for the transmission of Buddhism across the trans-Himalayan area, notably in Tibet and Bhutan.

The Swat Valley was ruled as an independent and peaceful state by an Islamic monarchy from 1914 to 1969, known as the Yusufzai State of Swat. More recently, Swat has been a centre of conflict, preceding and following the formation of a Taliban Emirate between 2008 and 2009. During those years, Swat was the scene of military operations aimed at thwarting the creation of a Taliban Emirate, which managed to control the district for a few months. The military operations were launched in the spring of 2009 and culminated in an exodus of about two million people into refugee camps set up in the regions surrounding Swat. What had once been a revered tourist destination for its archaeological and natural beauty became known worldwide as a location of intense conflict (Olivieri, 2017). Today, however, the Swat Valley has been pacified.

Barikot is the entrance town to the Swat Valley, with a population of c. 25,000 people. The modern town sits above the ancient city of Bazira, which in turn sits upon deposits reaching back to the Bronze Age of 1700 BCE. Bazira is where Alexander the Great constructed one of his fortresses (Figure 7.2).

7.2 The Italian Mission and Heritage Projects in Swat

The Italian Archaeological Mission in Pakistan has been active since 1955, particularly in the Swat Valley (see Olivieri, 2006; Olivieri *et al.*, 2021, Olivieri, forthcoming (b) for an overview of the history of Italian research in Pakistan). The Mission was established at the initiative of Giuseppe Tucci within the Italian Institute for the Middle and Far East (IsMEO). Between 1996 and 2011, it became a research unit of a new institution, the Italian Institute for Africa and the East (IsIAO), and then of the International Association for the Mediterranean and the East (ISMEO). Since 2020, the Italian Mission has worked in collaboration with the Ca' Foscari University of Venice and is now funded mainly by the Archaeology and Cultural Cooperation Division of the Italian Ministry of Foreign Affairs, by the Italian Ministry of University and Research, via ISMEO and by Ca' Foscari University of Venice.

Giuseppe Tucci's research during the 1930s and 1940s revealed the direct relationship between Swat and the development of Vajrayana Buddhism of Tibet, which later fostered digs at important Buddhist sites in the Swat Valley, some visited by Tibetan pilgrims. Excavations in ancient urban sites included those at Barikot and Udegram (the city of Bazira and Ora of the Alexandrographs), while early Islamic sites were also explored, such as the Ghaznavid mosque on Mount Raja Gira (11th century). The Italian Mission has now published over two thousand monographs, research papers, and archaeological reports (almost all in English), while also undertaking groundbreaking studies on palaeobotany, geoarchaeology, and ancient DNA. Since 1984, the Italian Mission has been excavating the ruins of the ancient city of Bazira located below the town of Barikot.

This chapter will describe two projects relating to Barikot/Bazira which were implemented directly by the Italian Mission: the 'Archaeology Community Tourism'

Figure 7.2 View of the acropolis of Barikot. © ISMEO.

(ACT) field school project and the 'Save the Archaeological Site of Bazira (Swat, Pakistan)' project, funded by the ALIPH foundation – the International Alliance for the Protection of Heritage in Conflict Areas. The ACT project began in 2011 and ran for five years with funding of around €2.5 million, supported by the Pakistan–Italian Debt Swap Agreement run by the Economic Affairs Division, Ministry of Finance (Government of Pakistan), and Italian Ministry of Foreign Affairs. The ALIPH project started in January 2022 and ran for 30 months with funding of over $470,000 from ALIPH and ISMEO. This chapter will also note a project to digitally document, conserve, and preserve Late Buddhist rock art reliefs funded by the British Council and delivered by EssaNoor Associates, a Pakistan firm of conservation architects and other partners, including the Italian Mission.

Prior to describing these projects, the chapter will consider the sustainability of heritage projects, the underlying principles of integration with sustainable development, and the progressive involvement of local communities as adopted by the Italian Mission in Pakistan.

7.3 Heritage, Sustainable Development, and Community Engagement

Heritage management has become of widespread interest to the development aid sector, being an important tool for constructing new types of economic activity such as heritage and religious tourism (e.g. Winter, 2008; Winter, 2009). Awareness has grown about the importance of building cultural identities for local communities and the need for sustainable approaches to heritage management. While the value of heritage projects, whether concerning restoration, conservation, or excavation, is easily recognised, there are many issues that must be considered before they are embarked upon, especially in post-conflict, rural, and socially conservative regions such as the Swat Valley.

Considerations When Devising a Heritage Project

Heritage projects can have a negative impact on their environment. When a horizontal or sub-horizontal surface is excavated, voids are created where they did not exist before, and new mounds arise from rubble or waste soil. Even the flaying of the original surface to remove vegetation cover can change the natural response of the soil or the built heritage to rain, wind, and other impacts. This can exacerbate erosion and surface run-off. Moreover, the presence of workmen can have an impact on local resources such as water, while also generating new quantities of waste and refuse.

Consideration must be given to how heritage projects might be perceived by the local communities of where they are undertaken. Restoration work of religious monuments that remain linked to ongoing practices may not meet with enthusiasm or favour by those communities living in the area today. In the Barikot area, for instance, the ACT project undertook the conservation of Buddhist stupa – mound-like structures that serve as places of meditation and are a symbol of the Buddha's mind, burial mound, and the path to enlightenment. Moreover, the exposure of the remains of past cultures may meet with disapproval and resistance from contemporary communities. At best, the community may simply view activities conducted by outsiders with suspicion, especially when the purpose and meaning are unclear to them.

Restoration, conservation, and excavation projects should make an educational contribution to the local community, which may involve increasing awareness of cultural heritage, its diversity, and international significance. This is sometimes referred to as the 'Third Mission'. Supporting responsible tourism can be an important tool for fostering cultural awareness; it can generate a sense of pride by the local community of their heritage while also providing direct economic benefits via income generation from tourism. Conversely, tourism may introduce elements of conflict if it is invasive or disrespectful of the social, cultural, or environmental situation.

Heritage-related projects can also have negative impacts on communities. Legal issues can arise when a heritage project is desirable on privately owned land and permission from the owner is required. If granted and the project is successful, it may seem appropriate for the site to be acquired by the state for further protection. This possibility may limit the willingness of private individuals to grant their land to archaeologists, especially if they fear their land will later be expropriated at non-market prices by the state.

Another consideration is the impact of heritage projects on the labour market. Project wages might be proposed that are higher than the standard wage, such as when a funding body imposes a minimum wage that is relatively high for the area. This can lead to conflicts in the short term and to false economic expectations among the employees for continuing payment at the higher rate once the heritage project is completed. A further issue is the potential increase in clandestine excavation activities, often linked to the illegal export of archaeological artefacts. Such activities are known to increase exponentially following short excavation projects. They can, however, be controlled, or decreased if already in existence, when the local community is involved with the project with recognised benefits.

Excavation projects necessarily involve a responsibility for the conservation and consolidation of the site after the end of the project. This requires an allocation of funds, often to employ caretakers who maintain responsibility for the excavation site on behalf of the projects. Excavation sites must also be left in a state that neither poses a risk nor causes damage to the environment, the local community, and their property. Another facet of legal responsibility concerns the need to involve the state in the notification of the sites to ensure they are legally protected. This step can shield the archaeological site from processes of encroachment or indiscriminate construction. If an archaeological site or area is notified, current legislation usually requires that a buffer zone is defined around it where permanent construction or mechanical earthworks are prohibited. Unfortunately, in the Swat Valley, new construction projects are sometimes launched in the vicinity of an archaeological site to prevent the possibility of further excavation and the possibility of expropriation by the state.

Sustainability by the Provision of Training

Well-planned and long-term projects enable the progressive inclusion of the local community into the restoration and archaeological work, eventually at the decision-making and even planning levels. This can be achieved through the sharing of techniques and knowledge, these being an essential means for attaining the long-term sustainability of heritage.

A reliance on externally sourced professional expertise for restoration and the use of non-local raw materials are unsustainable for the long term because they need continuous financial support. By training a local workforce and using local materials, the community can become active within the projects. External professionals are likely to be required in the first phase of a project, but these should make use of traditional techniques, which might be dry-stone and rammed earth masonry. Innovative technical elements should only be introduced when they can be realised with traditional techniques and locally available materials.

Once initial training has taken place, the contribution of external professionals will be to fine-tune and improve existing skills rather than to create new ones. Once this phase has passed, the external staff can be further reduced in number and guide the now-trained members of the local community in two directions. The first will be to undertake necessary restoration and excavation work themselves, and the second will be to act as trainers for further members of the community. In this way, it will be possible to almost completely remove the use of external professionals.

The result of this devolution of technical skill has two additional benefits. The first is that the skills can be applied by the local community in work other than heritage management and be used as bargaining weight in employment contracts. For this to occur, an important requirement is for publicly recognised certifications of the training to have been provided. The other result is an enhanced sense of ownership of their cultural heritage by the local community, this arising from their own contributions of manual and technical skill to its restoration and conservation.

This virtuous process used in restoration projects can also be used for archaeological excavations, as elucidated by Olivieri (2014; Olivieri et al., 2021). Under the best conditions, an excavation site can become a training ground. This is easiest achieved when large, open areas, rather than deep, narrow trenches, are excavated. This makes excavation slower and therefore more controllable, enabling the relationships between different parts of the site to be identified. In this way, already experienced workers can be used to manage the work of other less experienced or novice workers in specific areas. The work of each group can be controlled and monitored by the externally recruited professional archaeologists because each group is working in areas visible to the other. Excavation in deep, narrow trenches not only inhibits observation and control, but also divides the roles so that the workers undertake the physical work of excavation, while the professional archaeologists interpret the results.

Open-area excavations increase the awareness of the workers towards the depositional processes they are excavating. They may be able to contribute considerable local knowledge concerning soils, their compositions, and their post-process changes. The archaeologically recorded dwellings, floors, collapses, erosions, and washouts, maybe similar or recognisable to those occurring in the contemporary villages. This situation will diminish with the passage of time, because the rural world today is gradually being transformed. In this sense, the current historical moment in Pakistan is very important for this methodology of archaeological intervention. Even traditional knowledge related to dry stone or rammed earth masonry techniques will soon disappear, just as it disappeared 2–3 generations ago in most of the European countryside.

Under the best conditions, therefore, an excavation site can turn into a training ground. The Italian Mission projects have trained workers to become specialised in excavation, cleaning materials (ceramics in particular), flotation of organic materials, the maintenance of excavation areas, and as archaeological guides and specialised site keepers. The training also delivers a new awareness about cultural heritage, which generates social awareness and self-esteem. The benefit to the projects is a progressive reduction in the costs of external technical assistance on the budget. This has a multiplier effect in the case of future projects. Decreasing the extent of external professional assistance is also of great value for the cultural perception of the project by the communities, because there is often a preconception that funding mainly serves to pay the fees of external professionals rather than making a direct contribution to the local community.

7.4 The Archaeology Community Tourism (ACT) Field School Project

The Archaeology Community Tourism Field School (ACT-Field School) was funded under the Pakistani–Italian Debt Swap Agreement (PIDSA). PIDSA is run by the Italian Ministry of Foreign Affairs and Italian Cooperation (today AICS), jointly with the Pakistani Ministry of Finance and Economic Affairs Division (EAD). The project was delivered by the Italian Archaeological Mission in Pakistan jointly with the Pakistani Department of Archaeology and Museums (DOAM Federal). Soon after its start, the 18th Constitutional Amendment gave care of archaeological sites to the provincial level, causing the ACT project to be co-implemented with the Provincial DOAM of Khyber Pakhtunkhwa. It started in 2011 and was completed in 2016.

The aim of the ACT project was to develop a series of community-based engagement programmes characterised not only by the restoration, conservation, and excavation of various archaeological sites in Swat but also by the development and promotion of sustainable ecotourism initiatives. The programme included:

- The training and employment of local workers and university students.
- The establishment of the Swat Archaeological Guides Association (SAGA).
- The creation of an archaeological mountain trail park (Biagioli *et al.*, 2015).
- The promotion of a community-based ecotourism joint venture between the ACT project and the Sustainable Tourism Foundation Pakistan (STFP).

The ACT project was active in 12 archaeological sites that underwent significant restoration campaigns and were opened to the public. These sites included a

Figure 7.3 Conservation of the Jahanabad Buddha by specialists from Italy. © ISMEO.

rock art location in Jahanabad, where there is a colossal sculpture of a meditating Buddha that was intentionally disfigured in an attack attributed to the Taliban in September 2007 (Figure 7.3) and the conservation of Buddhist stupa.

The project undertook the complete reconstruction of the Swat Museum with an earthquake-proof structure (Figure 7.4). The museum had been damaged by the Kashmir earthquake of 2005 and by a Taliban attack in February 2008 which killed more than 50 people The war against the Taliban caused an exodus of people from the Swat Valley. When the Swat refugees returned in 2009, many were supported by the ACT project, which aimed to facilitate the district's return to normalcy through job creation and promoting Swat's cultural heritage management, seen as a potential asset of a future tourist economy. This part of the ACT project was fully achieved, enabling Swat to become a major tourist destination, reaching levels far above those prior to the conflict. Whereas Swat's previous attraction as a tourist destination has been primarily linked to its natural beauty and climate, the valley is now also visited for its unique archaeological heritage.

One of the most important achievements of the project was to bring heritage issues closer to local awareness in a way that was unexpected given how the project had begun immediately after a Taliban insurgency. Together with the ACT project, several local associations with aims to conserve and protect cultural heritage have arisen, disseminating information through Pashto language channels on social media. These associations were a consequence of the impact and popularity that the ACT project had achieved.

Figure 7.4 Newly reconstructed Swat Museum with an earthquake-proof structure. © ISMEO.

The investment made by the ACT project into tourism at a time of conflict was risky but achieved positive outcomes. During the first three years of the project, the number of domestic tourists increased by 150% to well above the pre-conflict level, with particularly important growth relating to heritage tourism. The popularisation of Swat's heritage was facilitated by social media. In 2009–2011, this represented a new and exciting communication phenomenon in Pakistan, providing an important interface across the country's cultural boundaries. While the ACT project could not be directly active on social media for security reasons, it encouraged visitors and journalists to become so. Engaging with the mainstream press (national and international) was also a strategic aim, using this to promote the project results in a widespread and viral manner. The training of tour guides was crucial, especially when they organised themselves into an association (SAGA) which, despite many difficulties, remains active. A particular difficulty faced by the new guides was that of self-perception: moving from the status of wage labourers to that of self-employed workers and entrepreneurs. This required changing a self-identity that had been inherited for generations.

The investment in 'religious tourism' was a second risky but successful initiative by ACT. Considering the context of intolerance and religious-based insurgency from which Swat was emerging, no one could have expected Swat's Buddhist heritage to become an important destination for tourists within a decade. Many of these tourists have come from countries where Buddhism is practised or where it represents an important element of historical and national identity, such as China and those of SE Asia. ACT's contributions had involved the reconstruction of the Swat Museum and the restoration of the colossal Buddha in Jahanabad, the Taliban destruction of both locations having been extensively reported in the national and international press. Political developments in Asia were also critical to the resurgence of tourism, especially those related to the start of the China–Pakistan Economic Corridor, part of the so-called New Silk Road.

As of June 2024, religious tourism in the Swat Valley and more generally in the province of Khyber Pakhtunkhwa has become an integral driver of the domestic economy. Its further development has become an important part of provincial policy and responds to a growing demand year by year. This has resulted in a high level of economic investment by the provincial government in the acquisition of archaeological sites, unparalleled in the previous history of Pakistan.

The fostering of such investment is one of the most important outcomes of the ACT project. The DOAM of the Province (DOAM KP) has invested about $3 million to acquire all the archaeological areas where the ACT project has been active, including the Barikot site. Thanks to this important contribution, since 2019 the entire archaeological area of Barikot is now government property, enabling planning of the archaeological work that continues today. Thanks to the ACT project and the state investment, Barikot has become an important asset in the framework of the region's cultural heritage and is becoming a major centre of attraction for heritage tourism. Of course, this development must be seen in the context of the changes that Pakistan undertook during the ACT years (2011–2016). One crucial change was the devolution of powers over heritage, museums, and archaeological sites to the provincial level (18th Constitutional Amendment, 2011), which altered the balance of power in government budgets in favour of a traditionally penalised sector.

The lessons of the ACT project have been taken up by several donors. These include the Prince Claus Foundation, which financed the restoration of wooden mosques in Swat, part of the largely late pre-modern Islamic architectural heritage. Also important is the inclusion of the ACT project's ethos in the World Bank's guidelines, which has financed a huge project on archaeological and religious tourism: KITE, the Khyber Pakhtunkhwa Integrated Tourism Development Project for Pakistan. KITE is aimed to support the province's cultural heritage with a budget of $77 million, mostly devoted to building light infrastructures at the sites and a promotional campaign for the archaeological riches of the province.

The Italian Mission has contributed to the design of two further projects, presented to the Italian Agency for Development Cooperation (an agency of the Italian government). The first is a Heritage Field School to be delivered by the Department of Archaeology and Museums of Khyber Pakhtunkhwa Province (DOAM KP), which is supported by soft loans (loans granted on soft terms with a long grace period for development projects). With funding of about €4 million, it envisages the establishment of five conservation laboratories within existing museum facilities, including one within the Swat Museum. It also includes a training plan for police and customs officers, which will be delivered by the specialised cultural heritage protection unit of the Italian Carabinieri.

The second project worth €3 million was designed by the Ca' Foscari University of Venice, the Italian Mission and the DOAM KP, and has been recently approved by the Italian Cooperation Program (AICS). This project has four aims: (1) to enhance seven archaeological sites from Swat, (2) to undertake pilot studies on the environmental impact of the sites under examination, (3) to assess the economic impact on baseline indicators of their tourism potential, and (4) to create Technical and Vocational Education and Training (TVET) opportunities.

The TVET will help to counter unemployment in rural areas and make trained local personnel available to archaeological and restoration projects. Without their contribution, it will be impossible for the provincial authorities to manage the large number of archaeological sites in the province. At present, a large part of the personnel employed in restoration at archaeological areas in the Swat region, such as at Amluk-dara, Abbashaeb-china, and Udegram, were trained with the ACT project.

In the six-year interval between the ACT project (2011–2016) and the starting of the ALIPH project (2022–2024), the Italian Mission was able to launch a series of six-month excavation campaigns at the site of Barikot (ancient Bazira), where the ALIPH project would then start in 2022. These six-month campaigns brought the results of the ACT project to fruition, employing members of the local community in excavation, guarding and control activities, and launching a series of excavation campaigns. The importance of the findings made Barikot the main archaeological site in the Khyber Pakhtunkhwa province.

This period encompassed the COVID-19 pandemic. The Mission team members made a scientific agreement with Exhalation Technology Ltd, a pharmaceutical industry-based company working in the field of biological markers, to monitor any signs of infections at the excavation/conservation site at Barikot. We developed a rapid testing device, CoronaCheck, which detects the presence of the SARS-CoV-2 virus in the breath of the subjects using exhaled breath condensate (EBC) and a bioenzymatic sensor. The device was extremely easy to handle, simple to use, and fully automated, capable of collecting and analysing the sample in a matter of minutes. After launching a vaccination awareness campaign that yielded excellent results among the worker communities, successful tests were carried out using the CoronaCheck device at the site, which all proved to be negative. The results were later confirmed by conventional tests. Thus, during the 2020–2021 excavation campaigns, the restoration and excavation sites were COVID-19-free environments, despite the large number of workers involved in the work.

7.5 The ALIPH and EssaNoor Projects

The ALIPH Foundation and ISMEO have funded the 'Save the Archaeological Site of Bazira (Swat, Pakistan)' project. This started in April 2022 and ended with a no-cost extension in December 2024. The project involves excavation and conservation work on the site's monumental acropolis in Barikot. It deals with colossal structures, two monumental terraces more than 20 metres high with retaining walls extending for more than 40–50 metres. The area of intervention extends approximately for two hectares on top of the hill that formed the acropolis of the ancient city. The conservation mainly deals with Shahi and Ghaznavid military and religious architecture, including the remains of a colossal podium that originally belonged to the Vaishnava temple built at the end of the 7th century on the ruins of a Buddhist sacred area. The entire acropolis was transformed into a fortified citadel by the Ghaznavids, at the time of early Islamisation of the region (11th century). It is a logistically complex area (there is no access to water, for example) requiring about 20 minutes of steep and strenuous walking from the road access to the archaeological site of the lower city.

The hill on which the lower city is located is the biggest landmark in the Swat Valley, overlooking its river. The project faced a problem of how to create a natural buffer zone from encroaching urbanisation. In agreement with the owners of the land (the hill is still private property, but has been leased to the Italian Mission since 1996), it was decided to graft the European olive (*Olea europaea*) onto the wild olive trees (*Olea cuspidata*, *Olea lanceolata*) that grow in thickets on the hill. This will both maintain the landscape with native plants and shield the archaeological site from the expanding nearby village and construction of a major highway to its immediate north.

The personnel who were selected for the ALIPH project were those already trained in the ACT project, but the scale of the intervention and the complexities involved had significantly increased. Specialised Italian restorers and occupational safety experts, as well as local engineers, trained the staff in an intervention and safety protocol, including the construction of scaffolding, which brought the local staff to a high degree of technical expertise (Figures 7.5 and 7.6). All the materials used were local, apart from a few imported chemical components required for minimal interventions. The stones were reused from collapse elsewhere, seeking to avoid any contribution to the destruction of the landscape that is taking place throughout Swat by the opening of quarries. The clay for the mortar, of excellent plastic quality, was obtained from the deepest sterile layers of the excavation carried out by the Italian Mission in the lower part of the ancient city. The brick dust, useful for its hydraulic qualities, was obtained at no cost from the waste from the brick furnaces. The sand and gravel were only taken from authorised quarries; the material for the mortar was made on site. Since no slaked lime is available in Swat, the site provided areas for slaking the lime and training the workers in the safety procedures related to that task. Practically, except for transport costs, all the materials for making the mortar used in the restoration and consolidation are local, obtained at low cost or from collapse and waste.

Once the large extensions of masonry had been completed, the personnel were trained by experienced rock climbers (the restorers themselves) in rope work to tackle the riskiest and highest precincts where scaffolding could not be erected. This procedure, which was completely new in the country, made it possible to further train the personnel, while also increasing training in occupational safety, which is one of the main and innovative aspects already initiated by the ACT project. All employed staff were insured against accidents at work during both the

Figure 7.5 Conservation of the Barikot acropolis. © ISMEO.

ACT and ALIPH projects, which is not usually the case for construction sites in Pakistan, other than for the staff of specialised companies.

While the ALIPH project was ongoing, some of the more technically trained members of staff were employed *pro tempore* on a project to digitise Late Buddhist rock art, funded by the British Council's Cultural Protection Fund (CPF) and

Figure 7.6 Detail of conservation of the Barikot acropolis. © ISMEO.

undertaken by EssaNoor Associates (April 2024–February 2025). The aim of this project is to digitally document, conserve, and preserve the Late Buddhist rock art reliefs of Swat, followed by those of the surrounding regions. This brief training interlude will be of great importance for the ALIPH staff, because when technical staff members return to the Barikot acropolis in the final phase of the work, they have the benefit of their new expertise.

The British Council/EssaNoor project also seeks to carry out community-based climate change adaptation training for the local community to better help them in taking care of the sites. Local community meetings have already been organised to raise public awareness, offer training opportunities, encourage active involvement, and emphasise the protection of the various rock art sites. Thanks to the efforts of the members of the Late Buddhist rock art digitisation project and the key logistical support and scientific expertise provided by the experienced members of the Italian Archaeological Mission in Pakistan (both local and external), several new rock art pieces have been identified during the currently ongoing field surveys to digitally document the already known specimens in Swat.

7.6 Reflections

This contribution cannot present an exhaustive summary of the work plans, methodologies, and results of the projects promoted by the Italian Archaeological Mission in Swat. It has emphasised how throughout the past 15 years, there has been a progressive inclusion of the local community and workers, which has been part of the Italian Mission's explicit work plan since 2011.

Prior to the ACT and ALIPH projects, the Italian Mission was aware of the special status it had achieved in the field of heritage protection and enhancement not only in Swat, but also elsewhere in Pakistan and beyond (e.g. Afghanistan). At the same time, it fully realised that this status imposed a further obligation, namely that of being able to make its work sustainable in the longer term. This process was initiated at a twofold level: that of governance, in agreement with Khyber Pakhtunkhwa Province's DOAM but above all at the local level, through associations, civil society, and the dissemination principles linked to the now diffused concept of 'Third Mission', which is aimed at contributing to growth of the local society by making a dedicated use of scientific knowledge. The aim was to include the local communities and organisations (universities, schools, press, civil society, associations) in the work of the specialists, to the point of making them active participants in the projects, rather than merely receptive to the results. We can call this new inclusiveness the 'Fourth Mission', which aims to include local society directly in the scientific and operational process by sharing roles and methodologies to create the conditions for long-term sustainability.

Overall, there has been a positive start to the process of entrusting greater amounts of heritage management to the community level. This process was started on an experimental basis with the ACT project and continued with the ALIPH-funded project on the Barikot acropolis, a project of greater technical, logistical, and scientific complexity. Finally, the British Council CPF-funded digitisation project of the Late Buddhist rock heritage of Swat sees the first results of a progressive inclusion of civil society and local communities in the direct management of the heritage, consistent with the responsibilities inherent to the integrated model of intervention on cultural heritage.

Acknowledgements

The list of thanks would be very long; I would like to acknowledge here all the individuals in charge of the teams of the projects mentioned (ACT, ALIPH, and British Council-CPF/EssaNoor), the organisations involved, starting with the DOAM KP and its director, Dr. Abdul Samad, the federal DOAM, and the two directors generals who succeeded one another during the period of the projects considered (Dr. Fazal Dad Kakar and Dr. Abdul Azeem), the AICS Pakistan and its directors, from Domenico Bruzzone to Francesco Zatta, obviously the Italian embassies in Pakistan and the Pakistani Embassy in Italy, their Heads of Mission, including the Consulate General in Milan and its consul generals. A great contribution, the greatest of which was certainly that of the local workers who accompanied the work and confirmed with their dedication the validity of the theoretical and procedural standlines, outlined above and applied at our sites.

References

Biagioli, C., De Chiara, M., Ferrari, E., Rana, A.R., Khan, S.A. and Vidale. M. (2015). *Living and Archaeological Landscapes of Swat. A Guide to Kandak and Kotah* Valleys. Lahore: Sang-e-Meel.

Olivieri, L.M. (forthcom.) 'Double-Crop Pocket Zones and the Empires. The Case of Swat', in: S. van Reden et al. (ed.) *Economies of the Edge: Frontier Zone Processes at Regional, Imperial, and Global Scales (300 BCE–300 CE)*. Heidelberg.

Olivieri, L.M. (2022). The Archaeology of Gandhāra. Oxford Research Encyclopaedia, *Anthropology*, pp. 1–19. DOI: https://doi.org/10.1093/acrefore/9780190854584.013.545

Olivieri, L.M. (2017). 'Archaeology from below in Swat, Pakistan: Heritage and Social Mobilization in a Post-Conflict Reality', in: P. Newson and R. Young (eds.) *Post-Conflict Archaeology and Cultural Heritage. Rebuilding Knowledge, Memory and Community from War-Damaged Material Culture*. New York: Routledge, pp. 217–237.

Olivieri, L.M. (2014). *Digging Up: Fieldwork guidelines for Archaeology Students*. Lahore: Sang-e-Meel.

Olivieri, L. M. (2006). Outline History of the IsIAO Italian Archaeological Mission in Pakistan (1956–2006), *East and West*, 4(1–3), pp. 23–41.

Olivieri, L.M., Filigenzi, A., Noori, N.A., Vidale, M. and Alberti, L. (2021). *Fieldwork Guidelines for Archaeology Officers* (Pashto, English and Dari editions). Kabul: UNESCO/ISMEO/Afghan Institute of Archaeology.

Winter, T. (2009). The Modernities of Heritage and Tourism: Interpretations of an Asian Future, *Journal of Heritage Tourism*, 4(2), pp. 105–115.

Winter, T. (2008). 'Asian Destination: Rethinking Material Culture', in: T. Winter, P. Teo and T.C. Chang (eds.) *Asia on Tour: Exploring the Rise of Asian Tourism*. Abingdon: Routledge, pp. 52–66.

8 Initiating the Transformation of Spanish Colonial Era Archaeology of Cagayan Province, Philippines, into Cultural Heritage

Caroline Marie Q. Lising and Grace Barretto-Tesoro

8.1 Introduction

Cagayan Province in north-eastern Luzon Island in the Philippines has numerous archaeological sites from different time periods. Research has focused on earlier periods, spanning from 709,000 years ago to 2,000 years ago. On the western side of the province along the border with Kalinga Province, sediments have abundant fossil fauna and stone tools dating to 709 kya (Ingicco *et al.*, 2018). The earliest known evidence for the exploitation of rice in the Philippine Islands is also found in this region. Across the valley on the eastern side of the province is the Callao Limestone Formation in Peñablanca town where fossils classified as *Homo luzonensis* were found within a cave that has become a top tourist attraction (Detroit *et al.*, 2019). Within this same karst formation, drawings dating to 3500 years ago have been documented in 11 caves (Jalandoni *et al.*, 2021). In the northern town of Lal-lo, shell middens with archaeological materials dating to 2200 cal BC and 1000 cal BC have been studied since the 1990s (de la Torre, 2000; Ogawa, 2000; Amano *et al.*, 2013).

In contrast to the prehistoric sites, those from the Spanish Colonial era have received little attention. A systematic survey of Spanish-era church ruins in Cagayan Province was undertaken in August 2022 as part of the Pantropocene Project, hosted at the Max Planck Institute of Geoanthropology, which is exploring the pre-colonial and colonial impacts in tropical forests throughout the former Spanish Empire, with a focus on the Philippine Archipelago. The survey was conducted in collaboration with the Cagayan Museum and Historical Research Center, alongside municipal government units, and recorded 21 sites including church structures, church complexes, and kilns, assigning them National Museum Codes (Figure 8.1). Although these sites had been documented in earlier cultural heritage initiatives, this is the first time they are being investigated from an archaeological lens (Figure 8.2). The survey underscored the importance of these sites, not only within the scope of the Pantropocene Project, but also for a broader archaeological understanding of Cagayan (Barretto-Tesoro *et al.*, 2023).

DOI: 10.4324/9781003491071-9

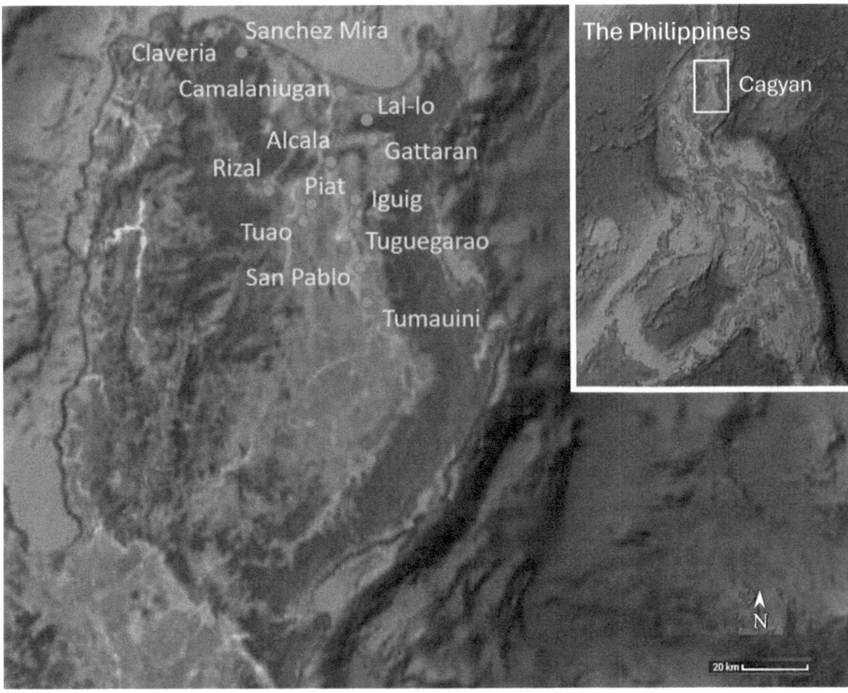

Figure 8.1 Map of the Philippines showing the archaeological sites and Spanish-era churches referred to in this chapter. © Mylene Quinto Lising.

8.2 Spanish Colonial Archaeology

Archaeological investigations are planned to begin exploring the Spanish colonial era in Cagayan. This initiative will signal a pivotal shift in archaeological scholarship, starting a new chapter in the province's historical exploration by focusing on the understudied aspects of its colonial heritage. Unlike previous projects in the region, the new excavations aim to transform archaeological data into culturally relevant heritage, with a specific emphasis on engaging local stakeholders. That aim is ingrained within the philosophical framework and execution strategy of the project by involving members of stakeholder groups from the project's inception and developing a collaborative approach with the participation of the Cagayan Museum, local and provincial governmental bodies, the Catholic Church, and educational institutions. We anticipate an expansion of stakeholder involvement as the project progresses, with their engagement deepening over time. The transformation of archaeological material into cultural heritage will be achieved when the value of such heritage has been co-created by those with vested interests in its preservation and interpretation.

Figure 8.2 Members of the Survey Team in 2022 with Cagayan local stakeholder representatives visit the Archbishop of Tuguegarao, Archbishop Ricardo Baccay (L–R: Architect Michael Tabao, President of the Cagayan Heritage Conservation Society; Dr. Leee Neri, School Secretary U.P. School of Archaeology; Kevin Baclig, Curator of the Cagayan Provincial Museum and Historical Research Center; Caroline Marie Quinto Lising of the Ateneo de Manila DSA; Dr Grace Barretto-Tesoro, Director of the U.P. School of Archaeology; Ja Lim of Tuguegarao City Information Office; and Mark Mabanag of the U.P.S.A.). © Mylene Quinto Lising.

Spanish colonial archaeology, falling under the Early Modern Period (1400–1830 CE) (Barretto-Tesoro and Acabado, 2023), is characterized by sites and artefacts closely linked with the arrival and presence of the Spanish in the Philippines. The most notable are churches (Figures 8.3 and 8.4), coins, and pottery from Europe and the Americas. During the Spanish colonial period of 1571 to 1898, European urban planning, evident from open spaces known as plazas associated with churches, elite residences, and government buildings, was implemented across the archipelago (Barretto-Tesoro and Acabado 2023).

In addition to these urban layouts, other significant structures include kilns (Figure 8.5; Barretto-Tesoro *et al.*, 2023; Neri *et al.*, 2020), lighthouses (Castillo-Noche, 2005), bridges (Castillo-Noche, 2011), roadways, loading docks, and cemeteries (Barretto-Tesoro, 2016). Military architecture, such as forts (Gatbonton, 1985; Javellana, 1997; Neri *et al.*, 2023), and sites associated with shipbuilding activities (Bolunia, 2013) were also developed. Furthermore, shipwrecks, including the San

Figure 8.3 Façade of the 17th-century Baroque Roman Catholic Malaueg Church, Rizal, under the jurisdiction of the Archdiocese of Tuguegarao. © Mylene Quinto Lising.

Diego (Desroches *et al.*, 1996), provide additional insight into the maritime history of the period. While European and American goods have been documented for this period (Barretto-Tesoro, 2021), many artefacts, especially ceramics from China, Vietnam, and Thailand, continued to be present. A distinctive pottery type from this era is a locally kiln-fired earthenware known as Manila ware, having European forms and designs (Arriola, 2010; Gatbonton, 2003), displaying the incorporation of foreign influences under Spanish colonial rule.

It is crucial to recognize that not all archaeological sites dating to the Spanish colonial era were under Spanish control (Acabado, 2017). Furthermore, local populations did not passively accept the changes introduced by the Spanish; rather, they exercised agency in response to colonialism such as through the practice of rituals (Lapeña and Acabado, 2017) or engaging with indigenous social networks (Acabado and Barretto-Tesoro, 2020). Current research explores floral and faunal evidence introduced during the Spanish occupation to contribute to a deeper understanding of the impact of Spanish colonization on the environment (Amano *et al.*, 2020).

Figure 8.4 The survey team at the Tocolana Church ruins in Lal-lo. The Tocolana ruins, formerly Iglesia San Vicente Ferrer de Tocolana, were established as a Spanish mission in 1604 and later incorporated into the parishes of Camalaniugan and Lal-lo. © Mylene Quinto Lising.

The churches, kilns, and ruins in Cagayan (Barretto-Tesoro *et al.*, 2023) were constructed during the Spanish colonial period. While many of the churches remain active and have religious functions, the kilns and ruins have been designated heritage sites and protected under national and local conservation efforts. We plan to excavate churchyards and their adjacent areas as part of the Pantropocene Project. Numerous pre-16th-century AD sites have been documented along the Cagayan River, where Spanish-established towns were later founded. We anticipate recovering materials that will elucidate the effects of colonization on land use.

Figure 8.5 The survey team inside the kiln (horno) in Camalaniugan, Cagayan. The kiln overlooks the Cagayan River and is considered the tallest and best-preserved in the region despite recent alterations. © Mylene Quinto Lising.

8.3 Cultural Heritage and Tourism

The deliberate integration of cultural heritage into tourism is a relatively recent initiative in the Philippines. In her address on 26 August 2021 at the Cagayan Heritage Conversations Series, former Secretary of Tourism Bernadette Romulo-Puyat highlighted how the COVID-19 pandemic prompted the Department of Tourism (DOT) to recalibrate its strategies. The shift emphasized sustainable and inclusive tourism, placing Filipino identity and heritage at the core of efforts to revitalize the sector. Romulo-Puyat stressed the importance of local communities in managing their cultural assets to ensure authentic representation, fostering cultural preservation, and enhancing tourist experiences. This approach aligns with global trends that prioritize meaningful, immersive experiences, allowing both domestic and international tourists to engage deeply with Filipino traditions, arts, and rituals.

Romulo-Puyat also acknowledged the United Nations World Tourism Organization's recognition of cultural and heritage tourism as key drivers of global travel. This form of tourism fosters cross-cultural understanding and tolerance, contributing to a more peaceful world. In the Philippine context, where regional diversity is celebrated, heritage tourism offers a powerful platform to unite the nation while respecting local differences. It allows Filipinos to reconnect with their history, thereby deepening national unity and pride.

Looking ahead, Romulo-Puyat emphasized that the future of Philippine tourism lies in balancing cultural preservation with sustainable tourism. The focus must extend beyond storytelling to actively involve communities in curating and managing their heritage. This collaborative approach will ensure that cultural heritage remains a dynamic part of community life, empowering the country to manage its past while building a culturally rich and vibrant future. In this way, heritage tourism not only contributes to economic development but also plays a critical role in strengthening the cultural identity of the Filipino people.

On 15 March 2023, newly appointed Secretary of Tourism Christina Frasco outlined the Philippines' ambitious goal to emerge as a tourism powerhouse in Asia within the next five years. Speaking at the launch of the 2023–2028 National Tourism Development Plan (NTDP) during a summit in Manila, Frasco emphasized the Department of Tourism's strategy to build a tourism industry deeply rooted in Filipino culture, heritage, and identity. This vision seeks not only to boost the country's global competitiveness but also to ensure that tourism is sustainable and resilient in the face of future challenges.

The key focus of the NTDP is on improving tourism infrastructure and accessibility to enhance the overall travel experience. Secretary Frasco highlighted that the transformation of the tourism industry requires a bottom-up approach that supports both well-established tourist destinations and lesser-known, emerging spots. The aim is to create a balanced tourism landscape that promotes inclusivity and provides opportunities for regional development.

Frasco also acknowledged the challenges faced by different regions and stressed the need to address these obstacles head-on to achieve lasting progress. By tackling these persistent issues, the DOT aims to foster growth not only through promotional campaigns but also by strengthening the tourism sector from the ground up. This comprehensive approach, Frasco asserted, is essential for the Philippines to achieve its vision of becoming a premier destination in Asia, while simultaneously preserving its cultural and natural heritage for future generations.

While the Department of Tourism's commitment to integrating cultural heritage into tourism development is encouraging, implementation of this goal faces several challenges. The integration of cultural heritage into tourism not only requires high-level policy commitments but also necessitates the development of local capacities. From first-hand experience, it is evident that many local governments in the Philippines lack dedicated cultural officers. Instead, cultural and heritage responsibilities are typically assigned to municipal tourism officers, whose primary roles often centre on tourism promotion rather than heritage management. As such, there is a pressing need for local governments to appoint specialized cultural officers who can focus on preserving and promoting their region's historical and cultural assets. Without such roles, the lofty objectives set by the DOT may struggle to materialize at the grassroots level.

Moving forward, the success of these policies will largely depend on how well local communities and governments embrace cultural heritage as a central element of their tourism strategies. Encouraging heritage appreciation through educational programs and promoting heritage sites for their historical significance, rather than solely for religious purposes, could significantly enrich the tourism experience and foster a deeper connection to the Philippines' rich cultural history.

Spanish-era heritage structures, particularly churches, attract visitors primarily for religious purposes. These churches are most frequently visited during religious observances, such as the Catholic traditions of Holy Week's "Visita Iglesia" and Christmas. These visits are often motivated by spiritual significance rather than a desire for historical, cultural, or archaeological learning experiences. Despite this, the potential for these heritage sites to serve as cultural education hubs remains largely untapped.

8.4 The Development of Cultural Heritage Management in the Philippines

The management of cultural heritage involves a multifaceted process that encompasses the conservation and presentation of archaeological sites, artefacts, and landscapes. Central to this process is the engagement of stakeholders from diverse backgrounds, each contributing unique perspectives and interests. The question of stakeholder engagement methodologies is a prominent issue in cultural heritage management discourse, with debates revolving around whether decision-making should be top-down, led by experts, or whether local communities should be empowered to manage their own. Lising's (2016) review of existing literature on cultural heritage management identified the value of inclusive practices and the empowerment of local communities (Barthel-Bouchier, 2013; Blake, 2000; Carbonei *et al.*, 2012; Flatman, 2012; Greene and Moore, 2010; Harrison, 2013;

Hatton and McManamon, 2000; Silberman, 2010; Smith, 2006; Smygin, 2002; Willems, 2008).

Cultural heritage management serves as a vital bridge between archaeological data and public relevance. Collaboration across disciplines transforms raw data into coherent narratives, enriching archaeological research with practical applications. This transformative process involves identifying values, facilitating exchanges, and co-creating knowledge to enrich archaeological understanding and promote broader stakeholder participation. Such an inclusive approach ensures sustainability and cultivates shared responsibility for preserving and promoting cultural heritage beyond academic realms (Acabado and Kuan, 2021; Acabado and Martin, 2022).

Effective management of cultural heritage, however, presents a complex challenge, requiring the establishment of a structured system overseen by trained professionals to ensure sustainable implementation. The absence of well-defined and organized systems for heritage management in the Philippines creates significant problems. Without clear frameworks and guidelines, management efforts become fragmented and inconsistent, leading to inefficiencies and gaps in preservation endeavours, and complicate the coordination and streamlining of management processes.

Archaeology's role in cultural heritage management is crucial. Many archaeologists lack formal training in management and communications, hindering broader engagement, and require further education for cultural heritage management. Unlike licensed professions, archaeology in the Philippines lacks standardized examinations, leading to varied competence levels. Despite its focus on material remains, archaeologists can broaden their scope by acquiring additional skills and collaborating across disciplines. Cultural heritage management demands competence, collaboration, and public engagement. Archaeologists face the choice of remaining traditional researchers or embracing a broader role in cultural heritage management.

The Peñablanca prehistoric archaeological cave sites in the south-eastern side of Cagayan Province illustrate the complexities inherent in cultural heritage management. Historically, these sites were administered through a top-down approach, characterized by the state authority's centralized control over decision-making processes (Lising, 2016). This approach frequently marginalized local communities, neglecting their valuable contributions to heritage conservation efforts. In recent years, there has been a growing acknowledgement of the necessity to engage local stakeholders in the management of community-based tourism projects. Collaborative research initiatives are being explored as promising avenues for fostering a more inclusive approach to heritage management as exemplified in practice by the Peñablanca Rock Art (Jalandoni *et al.*, 2021).

In a significant development, the National Museum of the Philippines (NMP) attained membership of the Protected Area Management Board of the Peñablanca Karst Formation, home to the Callao Cave. Additionally, the NMP has been entrusted with the management of this cave which had been designated as

Important Cultural Property in 2020. This long-awaited involvement of the NMP has effectively resolved jurisdictional disputes between the Provincial Government of Cagayan and the Municipal Government of Peñablanca. These conflicts had resulted in the closure of the cave to the public during the pandemic, with its eventual reopening to tourists on 18 October 2023.

When the *Homo luzonensis* research team resumed its excavation in Callao Cave on 5 May 2024, they were reminded of the necessity of interacting with and involving the local people and/or their representatives in their research, since concerns about cultural heritage management of the site have historically not been part of this research team's agenda in Callao (Lising, personal observation, 2015–2019). This will now be coordinated by the curator of the Cagayan Provincial Museum and Historical Research Centre who now sits on the Protected Area Management Board as a representative of the provincial government.

8.5 The Need for Training in Cultural Resource Management

Efforts are needed to develop tailored educational programs in the Philippines for cultural resource management (CRM) practitioners at both undergraduate and graduate levels. These programs should provide thorough training in heritage management principles, practices, and methodologies to equip practitioners with the necessary skills to excel in their roles. Additionally, consideration should be given to establishing a dedicated government agency or department tasked solely with overseeing CRM. A comprehensive review of the National Commission for Culture and the Arts (NCCA) mandate and structure could determine its suitability for this role, potentially serving as a central hub for heritage management activities nationwide. Encouraging provincial and local government units to establish structured units and allocate resources for heritage management is essential at the local level. Empowering local authorities fosters ownership of heritage management initiatives, leading to greater coordination and collaboration for more effective outcomes.

The introduction of mandatory CRM survey courses within relevant disciplines' curricula, along with interdisciplinary courses, would prepare future professionals by covering key concepts and methodologies. Furthermore, expanding course requirements to include fields like organizational development, public administration, and accounting may enhance the holistic approach to heritage management. By integrating interdisciplinary perspectives and skills, practitioners can address complex challenges and innovate solutions for cultural heritage preservation and celebration.

Effective cultural heritage management requires concerted efforts. Developing comprehensive educational programs, establishing dedicated oversight mechanisms, and empowering local authorities can enhance practitioners' capacity and ensure the preservation and promotion of cultural heritage for future generations. Recognizing parallels between cultural heritage management and ecosystems informs holistic and sustainable approaches to preservation and celebration.

8.6 The Importance of Stakeholder Engagement

Stakeholder engagement is fundamental to the cultural heritage management process, particularly in addressing potential conflicts arising from differing perspectives on heritage. Conservation and presentation phases of heritage management require interdisciplinary collaboration, involving archaeologists, historians, local communities, government agencies, and various interest groups. Blake (2000), Flatman (2012), and Greene and Moore (2010) emphasize the significance of engaging stakeholders throughout these phases to ensure that decisions consider diverse viewpoints and interests. Local communities often possess intimate knowledge of archaeological resources, shaped by cultural traditions and beliefs. Their involvement is crucial for fostering a sense of ownership and stewardship over heritage sites, thereby promoting their long-term preservation.

The shift towards a social perspective in heritage management recognizes the intrinsic connection between humans and their environment. David and Thomas (2008) highlight this shift, emphasizing the importance of including diverse perspectives in heritage management practices. However, despite international efforts towards inclusivity, some regions of the Philippines continue to adhere to top-down decision-making models (Heng *et al.*, 2020). For instance, the administration of archaeological sites, such as the Peñablanca prehistoric cave sites, often follows a hierarchical approach dominated by state authorities (Lising, 2016). This top-down approach undermines the agency of local communities and neglects their invaluable contributions to heritage management.

Efforts to empower local communities in cultural heritage management encounter numerous challenges, including bureaucratic barriers, resource constraints, and divergent stakeholder priorities. Furthermore, traditional conservation approaches have often prioritized scientific expertise over community perspectives, leading to conflicts and distrust (Kohl and McCool, 2016). To address these obstacles, a shift towards participatory approaches is imperative, whereby local communities are actively engaged in decision-making processes (Jones, 2017). By recognizing and incorporating local values and knowledge systems, heritage management practices can become more inclusive and responsive to community needs (Nissley and King, 2014; Waterton and Watson, 2015).

Moreover, the proficiency and expertise of individuals involved in heritage management have emerged as significant concerns. Many practitioners in the field lack adequate training and qualifications, resulting in suboptimal outcomes and missed opportunities for preservation and promotion. This deficiency in expertise is exacerbated by heritage management being relegated to a secondary priority, often approached opportunistically rather than with genuine commitment (Artioli, 2010; Boyd and Timothy, 2006; Cleere, 1989; Fairclough, 2002; Mcmanamon, 2000; Renfrew and Bahn, 2007). Consequently, heritage management initiatives may suffer from insufficient resources, support, and attention.

Inclusive cultural heritage management necessitates a fundamental shift towards participatory decision-making processes that prioritize community empowerment and collaboration. Scholars such as Cleere (1989), Fairclough (2002), and

Okamura (2010) advocate for the recognition and inclusion of diverse perspectives in heritage management discussions and implementation. Moreover, international charters, including the Burra Charter of 1981 and the UNESCO World Heritage Information Kit of 2008, underscore the pivotal role of community engagement in heritage conservation efforts.

However, translating these principles into practice requires comprehensive institutional reforms, capacity-building initiatives, and the cultivation of partnerships between government agencies, academic institutions, and local communities. The inconsistency in policies and their susceptibility to change with shifts in administrations pose significant challenges to heritage management progress. Alterations in government priorities and agendas may lead to changes in funding allocations or modifications to existing policies, resulting in uncertainty and instability within the heritage management landscape.

Nevertheless, it is crucial to acknowledge that adaptive changes can also yield benefits, enabling greater flexibility and responsiveness to the challenges in heritage management. By fostering a dynamic and adaptable approach, stakeholders can better navigate the complexities of heritage management, ensuring continued progress towards the preservation and promotion of cultural heritage for future generations.

8.7 Empowering Stakeholders for Spanish-Era Archaeology: Clergy, Local Leaders, Interest Groups, and Educational Institutions

The effective transformation of Cagayan Province's Spanish-era archaeology into culturally relevant heritage, appreciated by local communities, necessitates early and ongoing participation of diverse stakeholder representatives. This inclusive approach will ensure that stakeholders contribute meaningfully at various stages, preventing the entrenchment of gatekeepers and maintaining relevance.

Within the intricate tapestry of stakeholders involved in the preservation and management of Spanish-era church ruins and complexes in Cagayan, the Catholic Church emerges as a pivotal entity, represented by the esteemed Archdiocese of Tuguegarao, currently led by Archbishop Ricardo Baccay. As custodian of all properties housing these significant historical sites, the Catholic Church assumes a central role in shaping the trajectory of decisions concerning their conservation and governance.

The significant role played by the Catholic Church in the process of redefining archaeology as cultural heritage highlights a critical need for the implementation of comprehensive training and educational programs focused on cultural heritage management. This imperative is particularly pronounced among the ranks of the local clergy, especially parish priests situated in regions hosting significant archaeological sites. Among these clergy members, even individuals of esteemed stature such as retired archbishops are implicated in instances of inadvertent and sometimes reckless destruction of vital Spanish-era and World War II sites in Tuguegarao, often under the pretext of rebuilding or restoration endeavours. An example of this is when in the 1990s, the Archbishop Diosdado Talamayan bulldozed the bombed ruins of the convent of St. Peter's Cathedral in Tuguegarao to build a new building with commercial spaces available for rent. Another is that on 22 April 2024, the 17th-century Saint Ferdinand Parish Church in Ilagan City, Isabela, was destroyed by fire, caused by renovations to accommodate a new ceiling fan as the parish priest's first major project since his installation.

Expanding upon this observation underscores the urgency of equipping religious leaders with the knowledge and tools necessary for responsible stewardship of cultural heritage assets within their jurisdictions. Such training initiatives should encompass diverse facets of heritage management, including principles of conservation, historical preservation techniques, and protocols for site management and protection. By fostering a deeper understanding of the cultural and historical significance of religious sites, these programs can empower clergy members to become proactive advocates for their preservation and guardians against inadvertent damage or destruction.

Moreover, engaging with the clergy in educational endeavours presents an opportunity to cultivate collaborative partnerships between religious institutions and heritage conservation authorities. By fostering dialogue and cooperation, these partnerships can facilitate the development of mutually beneficial strategies for the sustainable management and interpretation of cultural heritage sites. Furthermore, the involvement of clergy members in cultural heritage management initiatives can extend beyond mere preservation efforts to encompass broader community engagement and educational outreach. Parish priests, in their role as community leaders, possess a unique platform to raise awareness and promote appreciation for local history and heritage among parishioners and the broader public. Through initiatives such as guided tours, educational workshops, and outreach programs, clergy members can play a pivotal role in fostering a sense of pride and ownership in cultural heritage assets, thereby contributing to their long-term conservation and sustainability.

Beyond the Catholic Church, an array of local government bodies constitutes another indispensable cadre of stakeholders, spanning the administrative hierarchy from village to provincial levels. These entities, spearheaded by elected officials such as village captains, mayors, and governors, collectively wield considerable influence in guiding policies and initiatives related to heritage preservation and management within their respective jurisdictions. Notably, within this framework, city, municipal, and provincial tourism officers assume multifaceted roles, often shouldering the additional responsibility of municipal cultural officers, particularly at the municipal level. Furthermore, the curator of the esteemed Cagayan Provincial Museum and Historical Research Center occupies a pivotal position as the de facto provincial cultural officer, lending invaluable expertise and strategic direction to heritage preservation endeavours.

In tandem with governmental bodies, local heritage interest groups, exemplified by organizations like the Cagayan Heritage Conservation Society, emerge as passionate advocates for the protection and promotion of cultural heritage assets within the region. Their grassroots efforts and advocacy play a critical role in galvanizing community support and mobilizing resources for heritage conservation initiatives.

Another integral part of the stakeholder ecosystem is the local educational institutions, including the University of St. Louis Tuguegarao and the Cagayan State University. These academic institutions contribute a wealth of expertise, research capabilities, and educational resources to heritage management endeavours, fostering interdisciplinary collaboration and knowledge exchange. Their involvement not only enriches the depth of scholarship but also cultivates a culture of community engagement and stewardship among future generations.

Thus, within this intricate web of stakeholders, characterized by diverse interests and expertise, collaborative partnerships and synergies are forged, laying the foundation for a holistic and sustainable approach to the preservation and promotion of Cagayan's rich cultural heritage.

The ongoing involvement of stakeholders throughout the transformation process is essential to its dynamism and relevance. As the process evolves, new challenges, opportunities, and priorities may emerge, requiring input from different stakeholders. By allowing stakeholders to enter the process at any stage as needed, the transformation remains adaptable and responsive to changing circumstances. This flexibility prevents protracted custodial dominance of individuals or groups who control access to resources or decision-making processes and promotes inclusivity and transparency.

Continuous training in skills, competence, and communication is vital for key representatives of stakeholder groups involved in the transformation process. As stewards of cultural heritage, these individuals play a critical role in guiding and shaping the process, ensuring that it aligns with best practices, ethical standards, and community needs. Training programs should cover a range of topics, including heritage conservation, community engagement, conflict resolution, project management, and communication strategies. By equipping stakeholders with the necessary knowledge, skills, and tools, training programs empower them to effectively contribute to the transformation process and advocate for the interests of their respective constituencies.

Sustained communication and collaboration among stakeholders are essential for fostering trust, building consensus, and overcoming challenges. Regular meetings, workshops, and forums provide opportunities for stakeholders to exchange ideas, share experiences, and address concerns collaboratively. Transparent communication channels, such as newsletters, websites, and social media platforms, facilitate the dissemination of information and updates about the transformation process, ensuring that stakeholders remain informed and engaged.

8.8 Proposed Excavation Project and Community Engagement

As the planning phase for the excavation of Spanish-era church complexes and ruins progresses, a proactive engagement with local stakeholders has been initiated. This initial involvement is anticipated to evolve into a more profound and multifaceted interaction once excavation activities commence. Beyond mere labour contributions, the participation of students from nearby educational institutions will enrich the excavation process, providing learning opportunities and fostering a sense of ownership among younger generations.

The public will always be welcomed to witness and engage with the excavation efforts, with an emphasis on encouraging questions and dialogue with archaeologists. This inclusive approach not only demystifies the archaeological process but also promotes a deeper understanding and appreciation of the significance of these historical sites.

To facilitate broader public awareness and understanding, forums will be convened periodically, where members of the archaeological research team will provide updates on progress, share insights into the importance of the sites, and outline the expected outcomes of the research. These forums will serve as platforms for dialogue and education, catering to diverse audiences and ensuring accessibility to information.

In addition to this, exhibitions showcasing the discoveries from the excavations will be organized, with active participation from local communities at various administrative levels. This collaborative approach not only highlights the tangible outcomes of the research but also strengthens community ties and fosters a sense of collective pride in local heritage. Moreover, the excavation experience will be structured as a comprehensive training program for local cultural workers and students interested in heritage management. By providing hands-on learning opportunities and mentorship from experienced professionals, this program aims to build local capacity and expertise in heritage conservation and management.

During periods when excavation activities are not ongoing, continuous engagement with local stakeholders will be maintained to further refine the research agenda and develop programs focused on awareness, conservation, interpretation, and presentation of cultural heritage. Additionally, opportunities for skills development and capacity building will be offered to ensure sustained community involvement and empowerment in the stewardship of cultural heritage assets.

8.9 Reflections

The effective transformation of Cagayan Province's Spanish-era archaeology into culturally relevant heritage requires the active participation and involvement of diverse stakeholder representatives from the outset of the process. Continuous training in skills, competence, and communication is essential for key stakeholders throughout the process to ensure their effective engagement and stewardship of cultural heritage. By fostering collaboration, inclusivity, and transparency, the transformation process can generate meaningful outcomes that resonate with local communities and contribute to the preservation and promotion of cultural heritage.

Empowering stakeholders, particularly local communities, is essential for the effective conservation and presentation of archaeological resources. By embracing inclusive practices and prioritizing community engagement, cultural heritage management can become more responsive, adaptive, and sustainable. Efforts to promote community empowerment must be accompanied by institutional reforms, capacity-building initiatives, training, and the fostering of partnerships between stakeholders. Ultimately, inclusive cultural heritage management ensures that heritage resources are preserved and interpreted in a manner that respects diverse perspectives and serves the needs of both present and future generations.

Acknowledgements

We are grateful for support from Manuel N. Mamba, the Governor of Cagayan and the following members of the Provincial Government: Atty. Charo Mamba-Villaflor (Provincial Administrator), Ms. Jeni Junio (Provincial Tourism Officer), Mr. Kevin Baclig (Curator, Cagayan Provincial Museum and Historical Research Center, CMHRC), Ms. Lorie Decena (CMHRC Deputy Curator) and Ms. Mia Baquiran (Cagayan Provincial Information Office). With regard to Tuguegarao City, we are grateful to Mayor Maila Ting-Que, Dr. Raul Ting, Ms. Ja Lim (City Tourism Office); Michael Tabao (Architect at the University of St. Louis Tuguegarao); and the Reverend Archbishop Ricardo Baccay. Mrs. Wilma Guzman kindly hosted the Survey Team at the beachside retreat centre of the Medical Colleges of Northern Philippines for our work at Pamplona. In Claveria we were supported by Vice Mayor Fredelino Agpuldo Gattaran, Vice Mayor Olive Pascual, and Ms. Bec Bautista (Municipal Tourism Officer), in Piat by Mayor Leonel Guzman; Mrs. Beatrice Guzman Villacete, and in Rizal by Mayor Joel Ruma. For survey in Tuao we thank Sangguniang Bayan Member Marcita Mamba Perez. Thank you to Dr. Patrick Roberts from the Max Planck Institute of Geoanthropology, and Dr. Leee Neri, Mark Mabanag, and Arcadio Pagulayan from UP School of Archaeology.

References

Acabado, S. (2017). The Archaeology of Pericolonialism: Responses of the "Unconquered" to Spanish Conquest and Colonialism in Ifugao, Philippines, *International Journal of Historical Archaeology,* 21(1), pp.1–26.

Acabado, S. and Barretto-Tesoro, G. (2020). 'Places, Landscapes, and Identity: Place Making in the Colonial Period Philippines', in: C.D. Beaule and J.G. Douglas (eds.) *The Global Spanish Empire: Five Hundred Years of Place Making and Pluralism*. Arizona: University of Arizona Press, pp. 200–221.

Acabado, S., and Kuan, D.-W. (2021). 'Indigenous Peoples: Heritage and Landscape in the Asia Pacific', in: S. Acabado and D.-W. Kuan (eds.) *Indigenous Peoples, Heritage and Landscape in the Asia Pacific: Knowledge Co-Production and Empowerment.* New York: Routledge, pp. 1–15.

Acabado, S. and Martin, M. (2022). *Indigenous Archaeology in the Philippines: Decolonizing Ifugao History.* Tucson: University of Arizona Press.

Amano, N., Bankoff, G., Findley, D.M., Barretto-Tesoro, G. and Roberts, P. (2020). Archaeological and Historical Insights into the Ecological Impacts of Pre-Colonial and Colonial Introductions into the Philippine Archipelago, *The Holocene*, 31(2), pp. 1–18.

Amano, N., Piper, P. J., Hung, H.-c. and Bellwood, P. (2013). Introduced Domestic Animals in the Neolithic and Metal Age of the Philippines: Evidence from Nagsabaran, Northern Luzon, *The Journal of Island and Coastal Archaeology*, 8, pp. 317–335.

Arriola, D.M. (2010). *Revisiting Manila Ware: Characterization of a Local Pottery*. Master's Thesis, Archaeological Studies Program, University of the Philippines.

Artioli, G. (2010). *Scientific Methods and Cultural Heritage: An Introduction to the Application of Materials Science to Archaeometry and Conservation Science. Oxford:* Oxford University Press.

Barretto-Tesoro, G. (2016). *Himlayan, Pantiyon, Kampo Santo, Sementeryo: Exploring Philippine Cemeteries*. Quezon City: University of the Philippines Press.

Barretto-Tesoro, G. (2021). *The Ruins of Pinagbayanan: A Photo-Essay on the Archaeology of a Late Nineteenth Century Philippine Town*. Quezon City: University of the Philippines Press.

Barretto-Tesoro, G. and Acabado, S.B. (2023). 'Incorporating Indigenous Perspectives in Chronology Building: Rejecting the Three-Age System in Philippine Archaeology', in: D. Docot, S.B. Acabado and C.C. Camposano (eds.) *Plural Entanglements*. Quezon City: Ateneo De Manila University Press, pp. 28–56.

Barretto-Tesoro, G., Mabanag, M., Baclig, K., Tabao, M., Lising, C.M.Q., Neri, L. and Roberts, P. (2023). Archaeological Survey in Cagayan Valley, Northern Philippines as part of the PANTROPOCENE Project, *Proceedings of the Society of Philippine Archaeologists*, 12, pp. 99–118.

Barthel-Bouchier, D. (2013). *Cultural Heritage and the Challenge of Sustainability*. New York: Routledge.

Blake, J. (2000). On Defining the Cultural Heritage, *International & Comparative Law Quarterly*, 49, pp. 61–85.

Bolunia, M.J.L.A. (2013). 'Astilleros: The Spanish-Period Shipyards of Donsol, Pilar, and Magallanes', in: J.N. Miksic and G.G. Yian (eds.) *Ancient Harbours in Southeast Asia*. Bangkok: SEAMEO SPAFA Regional Centre for Archaeology and Fine Arts, pp. 115–148.

Boyd, S. W. and Timothy, D. J. (2006). Heritage Tourism in the 21st Century: Valued Traditions and New Perspectives, *Journal of Heritage Tourism,* 1(1), pp.1–16.

Carbonei, F., Oosterbeek, L. and Costa, C. (2012). The Educational and Awareness Purposes of the Paideia Approach for Heritage Management, *Natural Hazards and Earth System Sciences*, 12, pp.1983–1986.

Castillo-Noche, M.M.L.D. (2005). *Lonely Sentinels of the Sea: The Spanish Lighthouses of the Philippines*. Manila: University of Santo Tomas Publishing House.

Castillo-Noche, M.M.L.D. (2011). *Puentes de Espana en las Filipinas: The Spanish Colonial Bridges in the Philippines*. Manila: University of Santo Tomas Publishing House.

Cleere, H. (1989). *Archaeological Heritage Management in the Modern World*. London: Unwin Hyman.

David, B. and Thomas, J. (2008). 'Landscape, Archaeology, and Dwelling', in: B. David and J. Thomas (eds.), *Manual of Landscape Archaeology*. Walnut Creek, CA: Left Coast Press, pp.300–6.

De la Torre, A. (2000). Preliminary Report on the Lal-lo, Cagayan Archaeological Project: Clemente Irigayen Property Site (II-1995-O), Sta. Maria, Lal-lo, Cagayan, *Journal of Southeast Asian Archaeology*, 20, pp. 67–110.

Desroches, J.-P., Casal, G. and Goddio, F. (1996). *Treasures of the San Diego*. Manila: National Museum.

Détroit, F., Mijares, A.S., Corny, J., Daver, G., Zanolli, C., Dizon, E., Robles, E., Grün, R. and Piper, P.J. (2019). A New Species of Homo from the Late Pleistocene of the Philippines, *Nature,* 568, pp. 181–186.

Fairclough, G. J. (2002). 'Cultural Landscape, Computers and Characterisation', in G. Burenhult (ed.) *Archaeological informatics: Pushing the envelope. Oxford:* Archaeopress, pp. 277–294.
Flatman, J. (2012). The Heritage Policy Group: Policy Development and Community Engagement at the UCL Institute of Archaeology, *Archaeology International*, 15 (1), pp.19–23.
Gatbonton, E.B. (1985). *Bastión de San Diego*. Manila: Intramuros Administration.
Gatbonton, E.B. (2003). 'Manila Ware ReExamined', in: C.O. Valdes (ed.) *Pang-Alay Ritual Pottery in Ancient Philippines*. Makati: Ayala Foundation, Inc. and Oriental Ceramic Society of the Philippines, Inc., pp. 75–77.
Greene, K. and Moore, T. (2010). *Archaeology: An introduction*. Abingdon: Routledge.
Harrison, R. (2013). *Heritage: Critical Approaches.* New York: Routledge.
Hatton, A. and McManamon, F. P. (2000). *Cultural Resource Management in Contemporary Society: Perspectives on Managing and Presenting the Past*. London: Routledge.
Heng, P., Stark, M. and Phon, S. (2020). De-Exoticizing Cambodia's Archaeology Through Community Engagement, *Journal of Community Archaeology & Heritage, 7* (1), pp.1–16.
Ingicco, T., van den Bergh, G.D., Jago-on, C., Bahain, J.-J., Chacón, M.G., Amano, N., Forestier, H., King, C., Manalo, K., Nomade, S., Pereira, A., Reyes, M.C., Sémah, A.-M., Shao, Q., Voinchet, P., Falguères, C., Albers, P.C.H., Lising, M., Lyras, G., Yurnaldi, D., Rochette, P., Bautista, A. and de Vos, J. (2018). Earliest Known Hominin Activity in the Philippines by 709 Thousand Years Ago, *Nature*, 557 (7704), pp. 233–237.
Jalandoni, A., Faylona, M.G.P., Sambo, A.S., Wills, M.D., Lising, C.M., Kottermair, M., Loreiga, X.E. and Taçon, P.S.C. (2021). First Directly Dated Rock Art in Southeast Asia and the Archaeological Implications, *Radiocarbon*, 63(3), pp. 925–933.
Javellana, R.B. (1997). *Fortress of Empire: Spanish Colonial Fortifications of the Philippines, 1565–1898*. Makati City: Bookmark.
Jones, S. (2017). Wrestling with the Social Value of Heritage: Problems, Dilemmas and Opportunities, *Journal of Community Archaeology & Heritage*, 4(1), pp. 21–34.
Kohl, J. M. and McCool, S. F. (2016). *The Future Has Other Plans: Planning Holistically to Conserve Natural and Cultural Heritage.* Arvada: Fulcrum Publishing.
Lapeña, Q.G. and Acabado, S.B. (2017). Resistance Through Rituals: The Role of Philippine "Native Pig" (Sus scrofa) in Ifugao Feasting and Socio-Political Organization, *Journal of Archaeological Science*, 13, pp. 583–594.
Lising, C.M. (2016). *The Cultural Heritage Management of the Prehistoric Archaeological Cave Sites in Peñablanca Cagayan Luzon Island, Philippines*. Unpublished MA Thesis. Archaeological Studies Program, University of the Philippines Diliman.
McManamon, F. P. (2000). 'The Protection of Archaeological Resources in the United States: Reconciling Preservation with Contemporary Society', in: A. Hatton and F.P. McManamon (eds.) *Cultural Resources Management in Contemporary Society: Perspectives on Managing and Presenting the Past*. London: Routledge, pp. 40–54.
Neri, L.A.M., Basilia, P.G.A. and Luga, E.A.C. (2020). The Alubijid Spanish Kiln: The Archaeology, Colonial History, and Cultural Significance, *Journal of Philippine Local History and Heritage*, 6(1), pp. 105–140.
Neri, L.A.M., Caballero, J.A., Guirgio, A., Chua, J.R. and Acasio. J.R.F. (2023). Probing the M'ranao-Chinese Relations Before 1898: The Archaeology of Lanao del Norte, *Chinese Studies Journal,* 17, pp. 5–39.

Nissley, J. and King, T. (2014). Creative Mitigation: Alternative Strategies for Resources, Stakeholders, and the Public, *Advances in Archaeological Practice,* 2(1), pp. 28–57.

Ogawa, H. (2000). 'Introduction: The Excavation of Lal-lo Shell Midden', in: H. Ogawa (ed.) *Excavation of Lal-lo Shell Midden.* Tokyo: Tokyo University of Foreign Studies, pp. 1–15.

Okamura, K. (2010). 'Introduction: New Perspectives in Global Public Archaeology', in: K. Okamura (ed.) *New Perspectives in Global Public Archaeology.* New York: Springer, pp. 1–9.

Renfrew, C., and Bahn, P. (2007). *Archaeology: Theories, Methods, and Practice* (6th ed.). London: Thames & Hudson.

Silberman, N. (2010). 'Beyond Theme Parks and Digitized Data: What Can Cultural Heritage Technologies Contribute to the Public Understanding of the Past?', in: M. R. N. (ed.) *Cultural Heritage and Technological Innovation.* Budapest: Bepress, pp. 1–13.

Smith, L. (2006). *Uses of Heritage.* Abingdon: Routledge.

Smygin, B. (2002). *A Variety of Conservation Principles and Methods—The basis of Contemporary Heritage Preservation Estrategias relativas al patrimonio cultural mundial. La salvaguarda en un mundo globalizado. Principios, practicas y perspectivas. 13th ICOMOS General Assembly and Scientific Symposium.* Madrid, Spain: Comité Nacional Español del ICOMOS, pp. 196–198.

Waterton, E. and Watson, S. (2015). *The Palgrave Handbook of Contemporary Heritage Research.* Palgrave Macmillan.

Willems, W. J. H. (2008). Archaeological Resource Management and Preservation: European Perspectives and Challenges, *Geoarchaeological and Bioarchaeological Studies*, 10, pp. 283–289.

Part II

Africa

9 Community and Archaeology at Amarna

Exploring Sustainable Heritage Strategies for Rural Egypt

Kate Spence, Anna Stevens, Gemma Tully, Fathy Awad and Hamada Kellawy

9.1 Introduction

Amarna is one of the most important archaeological sites in Egypt. Lying east of the River Nile in Middle Egypt in a wide desert bay bounded by cliffs to the north, east, and south, the archaeological site comprises the remains of a city built by the ancient Egyptian king Akhenaten as a state centre focused on the worship of the visible sun or Aten (Figure 9.1). Construction of the state centre, outlying religious buildings, and a sprawling settlement started sometime around 1340 BC, but the site was largely abandoned within 20 years of its inception. Amarna remains the largest and most complete settlement surviving from pharaonic Egypt. Because of its importance for the histories of religion and urbanism in the ancient world, the Egyptian government is developing a bid to have it placed on the UNESCO World Heritage List. At the same time, however, the site is under threat from environmental exposure, climate change, a rising water table, and expansion of adjacent towns, cemeteries, and fields.

The archaeological remains at Amarna are managed and protected by officials of the Egyptian Ministry of Tourism and Antiquities (MoTA), based in Cairo but devolved to regions and local sites and represented on the ground by antiquities inspectors, visitor centre staff, and site guards. Research teams work at the site via a permit system from the MoTA and Egyptian security forces. The site has seen a number of important excavation projects over the past 130 years, with the most notable early excavations being those of Flinders Petrie in 1891–1892, the work of the Deutsche Orient-Gesellschaft between 1907 and 1914, and that of the Egypt Exploration Society from 1921–1936. The Amarna Project/University of Cambridge has conducted surveys and excavations at Amarna since 1977 in cooperation with the MoTA and under the direction of our late colleague Professor Barry Kemp. Fathy Awad (Director, Southern Minia Antiquities) and Hamada Kellawy (Chief Inspector, Malawi Antiquities) are representatives of the MoTA working full time at Amarna and neighbouring sites, while Anna Stevens, Kate Spence, and Gemma Tully have all played an active role in fieldwork at the site. Fieldwork over the past five decades has focused partly on illuminating the lives of ordinary people

DOI: 10.4324/9781003491071-11

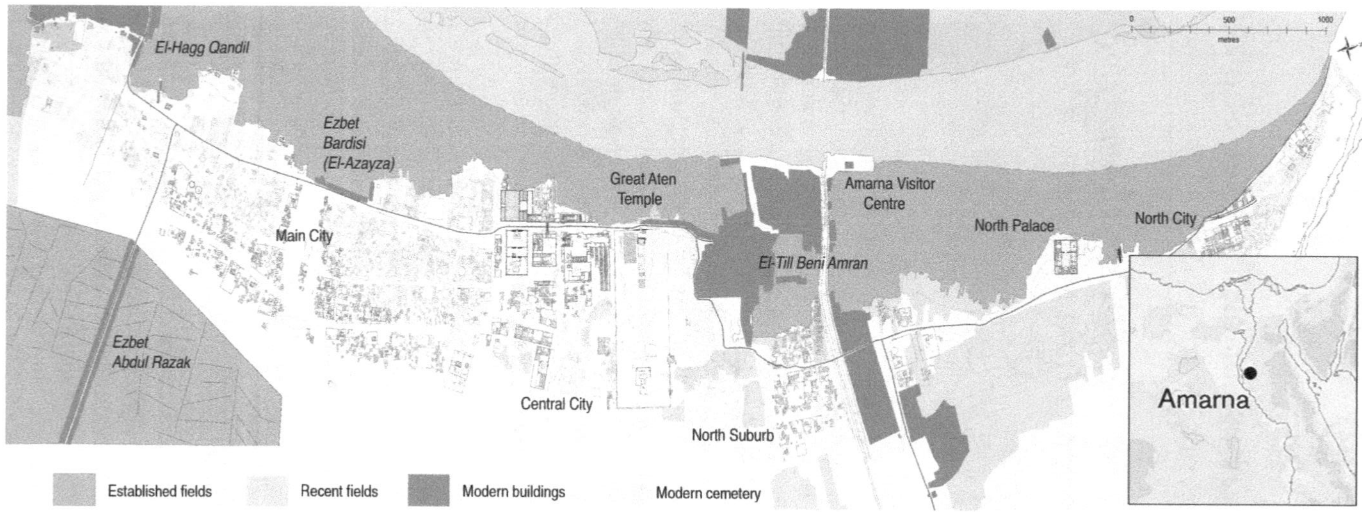

Figure 9.1 Map of Amarna, developed by Barry Kemp, showing the remains of the ancient city alongside modern villages, fields, and cemeteries. © The Amarna Project.

living at Amarna through excavation of housing, workshops, and cemeteries and partly on the royal ritual structures that formed the core of the city (Kemp, 2012). Since the 1980s, the Amarna Project has initiated conservation projects and site management proposals and partnered with the MoTA and Mallinson Architects on the construction of a Visitor Centre which opened in 2016.

The archaeological remains at Amarna are increasingly threatened by a range of factors. Environmental exposure is gradually eroding the delicate mudbrick walls left uncovered by early excavators, and climate change, including increased rainfall and a rising water table, is a growing threat. Looting is a reality at many sites in Egypt, particularly in times of unrest (Ikram, 2013). However, by far the greatest threat is population pressure and the need for land for agriculture, settlements, cemeteries, and industrial activity. In recent years, the three small villages adjacent to the archaeological site have gradually expanded into towns and land is needed for housing and cemeteries; work by Barry Kemp in the Great Aten Temple over the past decade has been designed in part to ensure that the archaeology of the structure is visible and recognisably important at a time when a very large proportion of that structure has been overbuilt by the expanding cemetery of the village of Et-Till. The use of mechanical irrigation and fertilisers allows crops to be grown in the former desert, and some parts of the site have now been lost to agricultural schemes (Figure 9.2). There has long been a disconnect between research teams and local communities, and little community understanding of what archaeologists are doing at the site or of the value and global importance of the remains on their doorsteps.

Amarna is some distance from major centres of tourism in Egypt and, despite its importance, has never attracted more than a tiny fraction of the international visitor numbers of sites such as Giza, Luxor, and Aswan. International tourist numbers at Amarna were highest in the late 1980s, peaking at around 17,000 per annum (David Phillips Associates, 1995, p. 14) of the c. 2 million tourist visitors to Egypt at the time (Steiner, 2006). At that time, the majority of visitors to Amarna were travelling between the major tourism centres of Cairo and Luxor along the Nile Valley by cruise ship or coach, stopping to visit Amarna for a couple of hours. From the late 1980s, the Egyptian state strongly promoted tourism with numbers rising rapidly to average over 10 million visitors per year in the 2010s. However, from 1989, terrorist attacks on tourists in the Nile Valley repeatedly affected tourism. Visitor numbers bounced back relatively rapidly in major centres following each event, and security provision and investment in infrastructure strongly focused on these key centres of tourism; however, visitors avoided the Nile Valley in the Minya Region of Middle Egypt because of security concerns, and, today there remains limited infrastructure in the area. As a result, only a few specialist tours and individual travellers frequent this part of Middle Egypt and international visitor numbers at Amarna remain well below their 1980s peak. In 2016 when the Visitor Centre first opened, there were 2484 Egyptian and international visitors to the site that year (school visits are not recorded in these numbers). Thus, while the expansion of tourism has brought significant economic growth and job opportunities to major tourist sites in Egypt, benefits have not been felt by local communities at Amarna.

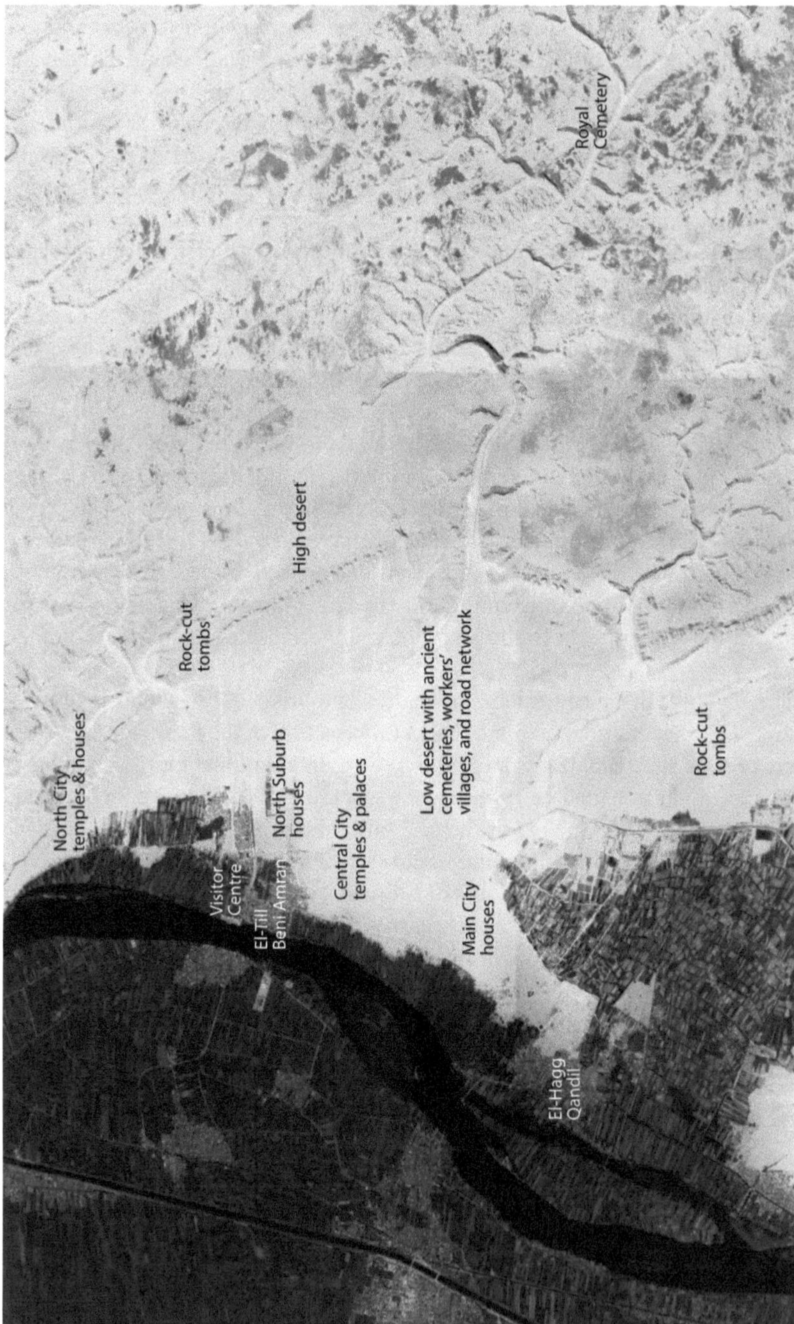

Figure 9.2 Marked-up Google satellite image of Amarna. Local towns have grown up on the site and artificially irrigated agricultural schemes threaten the archaeological site. © The Amarna Project.

The project that we present here was designed to address the specific set of issues encountered at Amarna, some of which are more broadly applicable to heritage sites elsewhere in Egypt, particularly ancient settlements with exposed mudbrick ruins, unexcavated archaeological remains, and low tourist numbers. The primary aim was to work with local MoTA staff to engage local communities in the archaeology of the site and to explore their interests, alongside developing a site management plan and resources for use with communities in the Amarna Visitor Centre. We also sought to pivot the presentation approach at the site from one previously aimed primarily at international visitors to material appealing to local audiences in Middle Egypt. We were particularly interested in working with children from the villages and towns near Amarna and over the course of the project expanded our remit to work in parallel with UK school children, developing resources that can be used both in Egyptian schools and UK Key Stage 2 teaching (children aged 7–11), with a view to developing a more nuanced understanding of life in modern Egypt and the Middle East and North Africa region.

9.2 The Project as Planned

A project entitled *Delivering sustainable heritage strategies for rural Egypt: community and archaeology at Tell el-Amarna* was developed in partnership between the University of Cambridge/Amarna Project and the MoTA and funded principally by a British Council Institutional Links grant (ID 261861975) under the Newton–Mosharafa Fund partnership with additional funding from the University of Cambridge and others; it ran from 2017 to 2020.

The project initially had three main and overlapping aims. The first was to develop the Amarna Visitor Centre, a large and well-designed heritage facility fitted out with models, information boards, and a small amphitheatre, into a local heritage and community hub, building on initiatives that had already been launched by the local MoTA. The Visitor Centre was designed initially for outside visitors to Amarna, but has great potential to also serve local community needs. This component of the project was to be run by a team of heritage outreach and engagement specialists from the MoTA in Cairo. It involved the creation of a training programme in outreach for Visitor Centre staff, to lead into the development of outreach programmes for local communities at the Visitor Centre. The second goal was to develop a suite of outreach resources for visitors, local community members, and others, for use at the Visitor Centre and more widely, namely, a children's book, a guidebook, a children's film, and a set of visitor information panels. The project leads for these elements were the Amarna Project/Cambridge University team and MoTA inspectors at Amarna. The third aim was to develop a comprehensive site management plan for Amarna. The project was launched with a workshop at the Visitor Centre in April 2018, which enabled team members from the Amarna Project/Cambridge University, the MoTA in Cairo and local MoTA staff to connect for an initial planning session. The project ran until 2020, although outreach and engagement activities at the site continue as we seek new ways to strengthen community links and sustainability.

9.3 Community Consultation and Engagement

Community archaeology has been developing in Egypt since the end of the 20th century (see Keshk, 2023; Tully, 2016) when the Quseir Project proposed a broad methodology for collaboration between communities and excavation teams (Moser et al., 2002; Tully, 2009). Broader adoption of collaborative approaches has been slow in Egypt but has become more widespread over the last five years, largely as a result of successful outreach and public engagement initiatives from Egyptian museums and funding that became available after the 'Arab Spring' (Abd el-Gawad et al., 2024; Keshk and Amin, 2024). Community engagement was intended to underpin our entire project – as a means of creating space for dialogue between local community members and the project team, and to enable coproduction of outputs – but it proved to be one of the trickiest elements to develop effectively.

Middle Egypt can be a complicated place for non-Egyptian researchers and collaborators to work. Since the early 1990s, and the attacks on tourists at this time, there has been a strong police presence in the region. At Amarna, police are stationed at the Dig House whenever research teams are present and are required to accompany non-Egyptian team members when they leave the house. Over time, the day-to-day work programme of the archaeological team has become regimented and predictable to help accommodate the police escort; these systems also mean that opportunities for informal interactions with members of local communities can be quite limited. The nature of community outreach, which often demands a more flexible and organic schedule than fieldwork, puts it somewhat at odds with these established routines. Local communities are also unlikely to be comfortable discussing issues such as potentially illegal land use in the presence of police. These factors, and wider political circumstances in Egypt at the time, resulted in unease around the project's planned community outreach elements and significant changes had to be made to proposed activities. Workarounds and compromise were a feature of the project.

Widespread consultation had been planned to discuss how the local community felt connected to Amarna and their understanding of the ancient city, along with ideas for the joint production of books, film and panels, and the activities to be included in the children's book. A large-scale survey proved unfeasible, but the Cambridge team was still able to consult with members of the local communities connected with the archaeological work (Figure 9.3).

Our Egyptian partners from the MoTA at Amarna were able to run activities for local adults and children in the Visitor Centre, visit schools, and develop questionnaires to interrogate understanding of the site and local needs; they also used their well-established Facebook presence to engage with local communities. As a result, school visits went up sixfold by the end of the project, and 100% of those questioned said they would recommend the Visitor Centre to family and friends. Feedback from local children led to the co-production of the children's book and fed into the film. At a later stage, this work was linked to activities in UK schools as we also outline below. Filming itself, however, and trialling of activities for the children's book had to be relocated to Cairo where we were fortunate to

Figure 9.3 Discussing Ancient Egypt and Amarna with local children at the Visitor Centre. © The Amarna Project.

be able to work with the British School in Cairo and the Awlady Orphanage. The changes also affected some activities on the Egyptian-led side of the project. In particular, the training programme devised by MoTA specialists at the Visitor Centre could not go ahead. Additional projects within the Visitor Centre such as the provision of modern reproduction artefacts for children's handling sessions and engagement with more recent heritage had to be abandoned, although these have since been partly revived through initiatives like the Petrie Museum's *Tutankhamun the Boy* project (Tully, 2023; Tully et al., 2023).

9.4 Site Management Plan

A key output of the project was an up-to-date site management plan which sets out the nature of Amarna's archaeological record, its excavation history, and the histories of the communities living around the site, as well as describing current and future threats to the site (Tully et al., 2020a). It also outlines the current policy context and develops a framework for action to conserve the site, drawing on existing management plans for key sites in Egypt and internationally to frame structure and content (e.g. Weeks and Heatherington, 2014; Simmonds and Thomas, 2015). The plan acknowledges the need for change and local development and attempts to mitigate future local conflict and to balance community access and economic needs with the importance of conservation, research, and interpretation at a site of global importance. The input of Fathy Awad, Hamada Kellawy, and their team through outreach activities was central to the development of the site management plan: their wide-ranging conversations with stakeholders and understanding of the local socio-political context, land use, and local needs of adjacent communities allowed a nuanced and balanced approach to future risk and its mitigation.

The site management plan was submitted to the MoTA in January 2020 and is available in both Arabic and English. It informs future activities at Amarna and underlies key aspects of funding applications. Most significantly it has fed into the development of the MoTA application for UNESCO World Heritage Status for Amarna. Preparation of the final dossier is ongoing, and members of the Amarna Project team have been invited to assist.

9.5 Site Information Panels

While the archaeology of Amarna is of exceptional importance, most of the archaeological remains are neither visually striking nor easy to interpret on the ground. Much of the site consists of low sandy mounds with low, eroded mud brick walls sometimes breaking through the desert surface. The absence of impressive stone monuments, so well attested and numerous at other archaeological sites in Egypt, renders the site (other than the rock-cut tombs) disappointing and incomprehensible to the majority of non-specialist tourists and does nothing to dispel local views that areas of the site are empty and unimportant.

In an attempt to make Amarna more comprehensible to all, we designed, produced, and installed, in August 2020, a set of 36 metal information panels around the site (Figure 9.4). These include text in Arabic and English, and visual information in the form of maps, plans, excavation photographs, and reconstruction

Figure 9.4 Site information panel outside the North Palace, one of few ancient buildings at Amarna with protective fencing. © The Amarna Project.

drawings of key structures in the city. The signs went through a process of formal review by the MoTA Scientific Publications Department before they were approved for printing.

The panels have been very well received, serving as useful illustrative tools for guides taking tourists around the city, but also interesting local residents such as those farming the fields nearby, clearly enlarging local knowledge of the site. Although we are concerned that the text on the panels is fading more quickly than expected in the bright sunlight, the interest in the panels and the rapid incorporation of them into guided tours of the ancient city suggest that this has been a successful intervention. The panels also serve, in part, as aides to help distinguish state-owned Antiquities Land from private land, addressing the problem that there are few visible markers (fences, survey posts, signs) to define the perimeter of the site. Several panels have been placed in open areas where the ruins are still partly buried and there is little immediately visible of the ancient buildings, with the goal of helping to demarcate these zones as part of the protected archaeological site; the panels provide a sense of formality, indicating that the archaeological land is important.

9.6 Guidebook

We also produced a guidebook, intended for Egyptian and international visitors to Amarna, and others with an interest in the site (Stevens, 2021). The book was designed to be both multivocal and to allow the reader to view Amarna with an 'archaeologist's eye' through contributions by researchers and heritage specialists involved in work at the site. Contributions were also included from local community members, on Amarna today and on local connections to the archaeological research. The guidebook explains the importance of Amarna and makes the results of recent archaeological research easily available to local communities as well as to archaeologists and guides and visitors to the site.

The guidebook was published by the American University in Cairo Press, as separate English and Arabic volumes. Authors' royalties from sales of the volumes are being donated to the Amarna Trust, a UK charity that supports work at Amarna, to support ongoing outreach programmes at the Amarna Visitor Centre. The project covered the printing costs of 1250 additional copies of the Arabic guidebook, which are being distributed free of charge at and around Amarna, including to cultural, educational, tourism, and other institutes in the wider Minia region. The volume has also proven valuable as an introductory text for university students in Egypt.

9.7 Education and Outreach

Education and outreach were primary foci of the project, but our planned activities were significantly affected by restrictions to the activities of the Cairo-led MoTA training scheme and limited opportunities to interact directly and brainstorm ideas with children on site at Amarna. There were also limitations on use of the Egyptian budget which required us to seek further funding for this aspect of the work.

As a result, we sought to pivot aspects of this outreach work towards a broader and more international collaboration, drawing on long-standing links between archaeologists from Cambridge and the communities at Amarna and widening our remit to engage with children in the UK. This aspect of the work was facilitated by a grant from the Cambridge University Arts, Humanities and Social Sciences Impact Fund. We were keen to explore ways of allowing children in Egypt and in the UK to find out a bit more about each other and to provide educational resources for both communities. Our aim was to foster interest in local heritage amongst children living around Amarna, and we were keen to investigate whether interest of other children might encourage local children to want to learn more about the archaeological remains on their doorsteps and to become ambassadors for their local heritage.

Considering potential benefits for UK schools, we noted that daily life in ancient Egypt is taught at Key Stage 2 (children aged 7–11), but teachers report limited available resources for daily life, with most information readily available to them being on mummification, death, kings, and religion (Doull, 2021). We aimed to develop more relevant teaching materials for use in schools, but also to enable Egyptian children to provide the answers to the sorts of questions UK children asked. Simultaneously, we hoped that finding out more about children in Egypt today would encourage a better understanding of the modern Middle East and North Africa in the UK and international context.

9.8 The Children's Book

Gemma Tully conducted workshops and activities with children at St Matthew's School and the Museum of Archaeology and Anthropology in Cambridge. While some discussions took place at Amarna, much of the outreach and engagement work in Egypt took place in Cairo at the British International School and Awlady Orphanage. Discussion and evaluation of a drawing exercise by University of Cambridge Museums' Evaluation Coordinator Sarah-Jane Harknett revealed that children in the UK knew very little about modern Egypt, with many assuming that it remained very similar to life in ancient Egypt, while Egyptian children were very keen to explain how different their lives are to those of people in the past (see insert box to film section).

Tully explored these differences with the groups she worked with, and the idea of parallel stories of ancient and modern Egyptian children living at Amarna emerged collaboratively, focusing on the lives of ancient Egyptian twins Nofret and Rahotep, and their modern counterparts, cousins Amira and Omar. The children introduce their lives to readers talking about their families and homes, clothing, food, work, transportation, and other themes.

This story has been published as a book, *Life Under the Sun. An Ancient Egyptian Story and Activity Book* (Tully et al., 2020b), and includes a series of activities for exploring life in ancient Egypt such as baking, mud brick making, a hieroglyph challenge, and weaving. The text of the book is in both Arabic and English. We felt it was important to use an Egyptian illustrator to ensure that the book was as appealing as possible to young Egyptian readers and that details of modern Egyptian life were correctly rendered; artist and illustrator Duaa El-Taweel provided wonderful evocative illustrations. The book was published in the Netherlands by BLKVLD and made available by them for international distribution; in Egypt, it was printed by Crest Press for local distribution. Approximately 1750–2000 copies have been distributed free of charge to local children at Amarna to date, and the Arabic version of the text is available for free download via the University of Cambridge website: https://www.arch.cam.ac.uk/research/projects/recently-completed-projects/life-ancient-egypt-amarna-resources-schools. Any profits from sales of the book are also directed to the Amarna Trust for ongoing outreach with schools by staff of the Amarna Visitor Centre.

Recent follow-up on site suggests that local children know the names of characters in the book suggesting that it has been widely used and we are considering producing further outreach resources based on the same characters. Testimonials received from teachers in the UK have also been very positive about the book.

9.9 The Film and Video Clips

In parallel with developing the children's book, we also produced a short film for use in schools and in the Amarna Visitor Centre. Directed by filmmaker Oliver Wilkins, *Under the Sun: New Ways of Understanding Amarna* has Arabic and English subtitles and is freely available on the University of Cambridge's YouTube channel: https://www.youtube.com/watch?v=KHTVPZWIznM&t=899s. Wilkins followed community archaeologist Gemma Tully as she explored the site at Amarna learning about daily life in ancient Egypt, talked with children in Cambridge about what they wanted to know about ancient Egypt, and asked archaeologists on site and children in Cairo to answer the questions raised in Cambridge. The film also follows the development of activities for the children's book. Particularly striking about the film is the obvious enthusiasm and excitement of the Egyptian children we worked with for the hands-on activities developed; such activities do not play a major role in Egyptian schooling, and the work suggests that this might be a fruitful direction to pursue in the future. The focus placed by Egyptian children on explaining modern life in Egypt as radically different from life in the past led to the focus on split-time presentation of our children's story and emphasises the need to listen carefully to local views and not to oversimplify in an enthusiasm to promote heritage. The children we worked with were interested in and proud of their ancient heritage whilst being extremely clear that things had changed, and they had no intention of being defined by it.

Comparing Ancient and Modern Egypt

(transcript of short composite film https://www.youtube.com/watch?v=kcpb53q9J5Y&list=PLqBlOsgK0Pvo-3sTFCmqtzGMYZ7FqrT0S&index=7)

Mostapha (Cairo):	A lot has changed. It's totally different to ancient Egypt, the way people think, the way they eat, everything they do…
Giselle (Cairo):	When I'm in England they ask me stuff like, do I go to school on a camel?
Mostapha:	Yes, I think many people think that we go to school on camels and that we dance like this [mimes cartoon ancient Egyptian dancing]. That stuff isn't true now.
Giselle:	I've never ridden a camel in my life.
Mostapha:	In modern Egypt everyone's thinking about the most famous footballer – Mohamed Salah.
Mansour (Cairo):	It's much more different than the ancient Egypt times – times of the pharaohs – it's much more advanced.

The dialogue between the children in Cairo and Cambridge about life in modern Egypt proved fascinating, and very revealing of the much broader interests of the children about life in other countries, but these proved impossible to fit within the length of the film. As a result, a series of additional clips of the communication were put together by Wilkins: https://www.youtube.com/playlist?list=PLqBlOsgK0Pvo-3sTFCmqtzGMYZ7FqrT0S.

Jokes

Charlotte (Cambridge):	Do you tell jokes today?
Saad (Cairo):	Children in Egypt love jokes.
Nour (Cairo):	I know a lot but I can't just think of them…
Saad:	Two bald men were fighting over a comb!
Charlotte:	Why did the banana go to the doctor… Because it wasn't peeling very well!
Nour:	No offence but Arabic jokes are more funnier than English!

Analysis by Sarah-Jane Harknett of drawing activities by children at St Matthew's School before and after workshops and activities organised by the project reveal a significantly enhanced understanding of daily life in modern Egypt, as well as a better understanding of archaeology and the role of archaeologists. The films are available for teachers to use in schools, but viewing and downloading figures have not been particularly high. That probably reflects the lack of promotion of the resources which were completed towards the end of the project and the need for better signposting to teachers.

The film seems not to have been much used in the Visitor Centre, and further investigation is needed to understand why this is the case. It is likely that the film would have been more accessible and of interest to audiences at Amarna had it involved people from the local community. Unfortunately, we had been unable to secure permission to involve the local community with this project. Technical issues in the Visitor Centre may also play a role in its limited use.

9.10 The Website

The *Teach Ancient Egypt Website* (www.teachancientegypt.com/about) was created by Gemma Tully and web designer Andrea Lorenz in an attempt to make the Key Stage 2 teaching content prepared for the children's book as outlined above more accessible and appealing to both teachers and pupils. The website was funded through a grant to Spence from Cambridge University's Arts, Humanities and Social Sciences Impact Fund. The website is in Arabic and English, and following consultation with both UK and Egyptian teachers of 7–11 year olds was designed to be visually appealing and cross-curricular: the beautifully illustrated activities developed from the children's book cover maths, art, science, literacy, and so on. While the website offers daily life focused learning resources that teachers say they need, and which move beyond the stereotyped content found elsewhere online, visits to the site remain low. Again, more work is needed here to enhance the site's discoverability and to promote it in the UK, Egypt, and the wider world.

9.11 Reflections from Amarna *By Fathy Awad and Hamada Kellawy*

Local reaction to both the children's book and guidebook has been very positive overall. As we would expect, children have a range of responses to the children's book – some are interested to sit and read it through, and others will just take a quick look. There might be ways of developing activities around the book in the future for children who are less interested in reading, like crosswords and other puzzles to help teach about the Amarna period. One important outcome has been that the books provide an accessible source of reliable information about archaeology and pharaonic history. These are popular topics on social media, but there is also a lot of misinformation online. Up to now, people have been reliant on social media for information in Arabic. Now they can check answers to questions they have, or questions that they see online, in books that are in their own homes. The high-quality printing and binding of both publications are also important for the long-term survival of the books. Books are not common items in local households at Amarna, but quality and durability are always important considerations when books are acquired.

The visitor information panels have also been well received, especially by Egyptian and international tourists and local university groups. One side effect is that they do now require monitoring and maintenance, and this is an added responsibility for local authorities. Overall, however, they have been a very positive addition to the site.

The process of developing the site management plan was very beneficial. It was a chance to focus on identifying problems and brainstorming solutions and to formalise the extensive knowledge that local inspectors have about the site and its management. Now that it is finished, the site management plan is not regularly consulted locally as inspectors go about their work, but this is not really its purpose. Perhaps most importantly, the process of putting it together drew attention to the problem of the poor definition of site boundaries. Working with the Amarna Project/University of Cambridge, and with funding from the American Research Center in Egypt, we have since been able to complete two projects to build boundary walls at threatened areas of the site, and further work is planned. In practical terms, there remains the ongoing task of monitoring the status of Amarna's varied monuments and developing a long-term conservation plan for the tombs and mudbrick buildings as a supplement to the site management plan. The hiring of additional guards, prevention of encroachment, and improvement of facilities such as roads, pathways, and staircases are all priorities. We also hope to communicate with irrigation authorities around paving the water canals to reduce the impact of irrigation waters on the site.

In terms of directly measuring the impact of the British Council project relative to the goal of building community engagement with Amarna, the international team is somewhat limited in what they can do here, and for the local inspectors, it is difficult to find the time and resources to undertake any large-scale community survey and data gathering. There are, however, general signs of increased engagement. One example is the annual festival of Sham El Naseem, which marks the beginning of spring, when people will traditionally climb up to the top of the cliffs at Amarna to 'smell the breeze'. We have noticed recently that more people now want to stop partway up the cliffs to visit the ancient tombs. Last year during Sham El Naseem, we opened the tombs specially for the local community to encourage this. It is also not uncommon to see local people standing and reading the visitor information panels. These are positive signs that the British Council funded project and initiatives that have followed are having real impact.

Moving forward, it will probably remain difficult for international researchers at Amarna to work directly with local communities beyond workshops in the Visitor Centre. The problem of how to create economic benefit from heritage for local communities also remains and is unlikely to be easily resolved. Only a very small number of jobs can be directly created, either through the MoTA (e.g. as site guards) or through the excavations. We also need to be mindful of what responsibilities are placed on local MoTA staff (e.g. if Amarna were to be placed on the World Heritage List). However, the importance of heritage outreach has been clearly demonstrated in recent years, and there is strong local support for this to continue. It may be possible to better utilise the children's film in the Visitor Centre and to make another film with local people asking them to reflect on Amarna and their local history.

Staff at the Amarna Visitor Centre currently run workshops for children, to invite them to learn about the importance of the region and the history of this important archaeological site, through activities like writing, painting, and colouring. The workshops ask how the ancient Egyptians lived, what is the value in protecting their relics, how people outside Egypt see our monuments and why they come to Egypt, what benefits heritage can bring to local people and for the country, why it is important to be proud of our monuments, and what will happen if we do not protect these? Staff of the Visitor Centre make visits to local schools, facilitated by an arrangement with the South Minia Education Department. School groups also come to Amarna, and we ensure that local inspectors have the training to be able to receive and educate these groups. There are many other ways in which the Visitor Centre might continue to grow to support local outreach. For example, we might invite local community members to the centre on national and public celebrations to increase their connection and loyalty. Future themes to explore through outreach might include folklore and local history, or how to make ancient objects through modern methods, to build bridges between now and the past.

9.12 Reflections from Cambridge, UK

Here, we reflect very briefly on the project from the perspective of the UK academic and institutional context, with additional reference to the UK's Research Excellence Framework (REF), because this project formed the focus of one of Cambridge Department of Archaeology's four submitted Impact Case Studies in 2021. The REF is a process undertaken every seven years by UK Research and Innovation, a public body sponsored by the Government Department for Science, Innovation and Technology to coordinate and deliver research funding to universities. The REF measures the quality of research outputs by every university department in all subject areas. It requires submissions to include impact case studies about how research has contributed to social well-being and/or economic growth.

We would like to note in particular the extent to which this project built on very long-term collaboration. Without the existing networks and trust established between the Cambridge/Amarna Project team and MoTA staff at Amarna over decades, this work would have been far more difficult, and the results would have been less satisfactory. At a time when archaeological project funding in the UK rarely extends beyond a few years, the limitations of such short-term approaches need to be weighed. We also note that, while many sites in Egypt face similar challenges, each site has a different spatial and socio-political context and specific solutions must be achieved through direct dialogue with local stakeholders. At sites where research involves international teams, this dialogue often cannot take place directly between foreign archaeologists and local communities emphasising the importance of Egyptian-led outreach projects whilst acknowledging the significant time and input then required of busy local MoTA staff. It is also important to understand local heritage management priorities and preferences, such as building protective walls around particularly threatened features, even when these appear to conflict with more familiar heritage priorities and practice elsewhere.

We stress the importance of working with community archaeology specialists with experience of working with similar communities and the requirement to be flexible with project content and delivery. This can raise issues with funding bodies: we are exceptionally grateful for the understanding and flexibility shown by the British Council team managing the Institutional Links projects, and also we were able to access additional funding from the University of Cambridge when Cairo elements of the project did not reach fruition and activities needed to adapt. Financial compliance is not always straightforward when operating between institutions and countries with different regulatory frameworks. Awareness of this is valuable, and we particularly note the need to be vigilant that individuals are not blamed for issues arising at institutional level over which they have no control.

It is important that work has impact, but measurement of impact is an inexact science: understanding what to measure, how, and when is important. Here, the value of designing measurements and milestones into a project from the start is clear, and this requires understanding of both system and context. We were fortunate that Tully had worked previously on a REF impact case study in the UK in addition to her experience in Egypt, but considerable flexibility was needed to understand what could and could not be measured in the context of the Visitor Centre, and not all collaborators were willing to submit testimonials, for example. In unfortunate timing, we were also trying to collect and collate information during the COVID-19 pandemic which was far from ideal as visitor numbers fell away, work became difficult and impetus was lost. Conducting surveys and questionnaires often must be carried out by local staff who are already busy with their routine activities. In the future, we would rethink our methodologies for measuring impact and would, in particular, observe more closely the distribution and use of resources (such as the children's book). It is also worth noting that outputs usually emerge right at the end of projects when funding has been spent and impetus has moved on to the development of the next project; without human or financial resources to collect data, measurement stalls: as a result, more of our current understanding of impact is anecdotal than we would have hoped. In a similar vein, we produced educational resources which we hoped would be more widely used but, without effective strategies to advertise and promote these, the use of some outputs has been limited.

9.13 Conclusions

This was a rewarding but challenging project which we hope will have a long-term impact through the development of the UNESCO World Heritage List bid, the use of the various outputs, the close working partnership between local communities, the MoTA and the Amarna Project/University of Cambridge team, and other projects such as the Petrie Museum *Tutankhamun the Boy* project, which have built on our work. We plan to continue the work we have done, and aim to build on the 'branding' of the characters and illustrations in the children's book, which are now well known at Amarna, to develop other outputs and trial further coproduction of resources. We also need to find more effective ways of promoting the outputs.

The project provided good evidence for the value of listening and other methods of co-production of resources, and we would argue that the relative success of the various outputs reflects the depth of partnership. The children's books and resources work well in principle and were well received (even if they have not been as widely used as we had hoped) because they were developed in collaboration with the children for whom they were intended. The film, by contrast, could not be co-produced at Amarna because we could not get security clearance to do this; as a result, the finished product lacks resonance for local groups. Local perspectives often differ from those in urban metropolises, and views and local knowledge at Amarna may differ significantly from those in Cairo.

We note the important role played by the Visitor Centre at Amarna as a public space and as a neutral space for engagement. Increasingly, it has moved beyond its initial role in supporting outside visitors to become a valuable element of local infrastructure and an outreach and education hub for local communities. Sustainability remains a major issue, however. Staff are stretched, and the Visitor Centre's operational budget is limited. While school visits to the Visitor Centre have increased sixfold, and adult visitors to the site at Amarna more than tripled between 2016 and 2023, the absolute numbers remain very low (7528 adult visitors in 2023), and thus, the economic impact for local communities also remains low. We remain committed to investigating further possibilities, but are mindful of the additional strain such activities place on local staff. This work is only possible because of their dedication to the site and the communities in which they operate.

Acknowledgements

We thank the people of Amarna for their interest and engagement with this project. Thanks are also due to the late Professor Barry Kemp and the Ministry of Antiquities staff at Amarna for their support and facilitation of the project. We are particularly grateful to Dr Adel Okasha (Director of the Central Department for Middle Egypt), Mr Gamal Abu Bakr (Director of Middle Egypt Antiquities), and Mr Hasan Taher (Director of Minia Antiquities) and to the staff of the Archaeological Sites and Museums Development Department and the Scientific Publications Department in Cairo. In Cambridge, we would also like to thank the staff of St Matthew's School for their input and the staff of the Museum of Archaeology and Anthropology for facilitating school visits. Analysis of the project work of the children from St Matthew's School was undertaken by Sarah-Jane Harknett.

This work was supported by an Institutional Links grant, ID 261861975, under the Newton–Mosharafa Fund partnership awarded to Kate Spence and Yasmin el-Shazly. The grant was funded by the UK Department for Business, Energy and Industrial Strategy and the Science and Technology Development Fund in Egypt and delivered by the British Council. Additional support was provided by the University of Cambridge through the Arts, Humanities and Social Sciences Impact Fund, the McDonald Institute for Archaeological Research, and the Thomas Mulvey

Fund. The Amarna Trust (www.amarnatrust.com/) provided further funding. We are grateful to the managers of all these funding bodies for their support. Ethical approval was obtained for all activities via the Cambridge University Department of Archaeology's ethics panel.

References

Abd el-Gawad, H., Volk, S., Van de Ven, A. and Stevenson, A. (2024). 'From Ancient to Living Egypt: Centering Lived Experience in Teaching Egypt Within and Outside Museums', in: J. Thum, C. Walsh, L.M. Jiménez and L. Saladino Haney (eds.) *Teaching Ancient Egypt in Museums: Pedagogies in Practice*. London: Routledge, pp. 201–217.

David Philips Associates (1995). *Tell el-Amarna: A strategy for conservation and Development. Prepared for the Supreme Council of Antiquities by the Partnership for Environmental and Cultural Heritage*. Wenlock: David Philips Associates.

Doull, K. (2021). Teaching Ancient Egypt: Developing Subject Knowledge, Historical Association, *Primary History article*. www.history.org.uk/publications/resource/10103/teaching-ancient-egypt-developing-subject-knowled

Ikram, S. (2013). Cultural Heritage in Times of Crisis: The View from Egypt, *Journal of Eastern Mediterranean Archaeology and Heritage Studies*, 1, pp. 366–371.

Kemp, B. (2012). *The City of Akhenaten and Nefertiti: Amarna and Its People*. London: Thames and Hudson.

Keshk, F. (2023). Heritage Outreach in Egypt Today, *Archaeologies: Journal of the World Archaeological Congress*, 19, pp. 28–38.

Keshk, F. and Amin, S. (2024). 'The Current Scene of Museum Education in Egypt', in: J. Thum, C. Walsh, L.M. Jiménez and L. Saladino Haney (eds.) *Teaching Ancient Egypt in Museums: Pedagogies in Practice*. London: Routledge, pp. 71–81.

Moser, S., Glazier, D., Phillips, J.E., Nasser el Nemr, L., Saleh Mousa, M., Nasr Aiesh, R., Richardson, S., Conner, A., and Seymour, M. (2002). Transforming Archaeology Through Practice: Strategies for Collaborative Archaeology and the Community Archaeology Project at Quseir, Egypt, *World Archaeology*, 34, pp. 220–248.

Simmonds, S. and Thomas, B. (2015). 'Stonehenge, Avebury and Associated Sites World Heritage Site Management Plan 2015', in: E. Nichols and R. Tyson (eds.) Paris: UNESCO.

Steiner, C. (2006). Tourism, Poverty Reduction, and the Political Economy: Egyptian Perspectives on Tourisms Economic Benefits in a Semi-Rentier State, *Tourism and Hospitality Planning & Development*, 3(3), pp. 161–177.

Stevens, A. (2021). *Amarna: A guide to the ancient city of Akhetaten*. Cairo: AUC Press.

Tully, G. (2009). Ten Years on: The Community Archaeology Project Quseir, Egypt, *Treballs d'Arqueologia*, 15, pp. 63–78.

Tully, G. (2016). 'From Community Archaeology to Civilian Activism: The Journey of Cultural Resource Management Through Heritage Dialogue in Egypt', in: G. Tully and M. Ridges (eds.) *Collaborative Heritage Management*. Piscataway: Gorgias Press, pp. 187–214.

Tully, G. (2023). 'Community Engagement', in: A. Stevens, G.R. Dabbs, F. Balestra, G. Tully, S. Schiødt, P. Rose, A.J. Clapham, P. Docherty, A. Garnett and M. Bertram (eds.) Tell el-Amarna, 2022, *Journal of Egyptian Archaeology*, 109, pp. 112–114.

Tully, G., Garnett, A., Bryant, E. and Wilson, C. (2023). 'Tutankhamun the Boy' at the Petrie Museum: Rethinking Childhood in Ancient Egypt Through Collaboration, *Egyptian Archaeology*, 63, pp. 44–48.

Tully, G., Kellawy, H., Spence, K., Stevens, A., Hussein, H. and the children of Amarna (2020b). *Life Under the Sun: An Ancient Egyptian Story and Activity Book.* Zandvoort: BLKVLD.

Tully, G., Stevens, A., Kellawy, H., Spence, K., Kemp, B. and Reyad, F.A. (2020a). *Tell el-Amarna Site Management Plan* 2020, Unpublished.

Watkins, O. (2019). *Daily Life in Ancient Egypt: Comparing ancient and modern Egypt.* Cambridge Archaeology/Amarna Project film resource accessible at: https://www.youtube.com/watch?v=kcpb53q9J5Y&list=PLqBlOsgK0Pvo-3sTFCmqtzGMYZ7FqrT0S&index=8

Weeks, K.R. and Hetherington, N.J. (2014). *The Valley of the Kings: A Site Management Handbook.* Cairo: American University in Cairo Press.

10 Community-Engaged Archaeology at the UNESCO World Heritage Site of Jebel Barkal, Sudan

Rebecca Bradshaw, Tohamy Abulgasim, Suzanne Davis, Elmontaser Dafalla, Sami Elamin, El Hassan Ahmed Mohamed and Geoff Emberling

10.1 Introduction

As many archaeologists work towards less colonial and more collaborative field projects, engagement with local communities and their ideas about heritage, preservation, and tourism have become increasingly essential. Imposing Euro-American heritage narratives and ideas about site exploration, management, and protection is now rightly seen as neo-colonial, particularly in areas like Africa, where modern states have developed directly from colonised pasts (Schmidt and Pikirayi, 2016; González-Ruibal, 2017; Lane, 2017; Ndoro et al., 2018). The integration of communities in everything including archaeological research, education, training, and heritage management is now increasingly mainstream. Over the past four years, the present authors have worked within a team of multinational archaeologists and conservators at the site of Jebel Barkal in Sudan. We have engaged with local communities in an increasingly collaborative way, responding to these important post-colonial approaches.

The initial aim of this chapter was to provide a detailed narrative and critical appraisal of the work we have undertaken so far and what we hope to do in the future. However, recent years have gone from turbulent to horrifying in Sudan. The revolution against General Omar al-Bashir and the Islamist state began in 2018–2019 but struggled for survival. The civilian-led transitional government held on through the ravages of COVID-19 in 2020 only to be overthrown by a military coup in 2021, orchestrated by the Sudan Armed Forces (SAF) and a paramilitary group, the Rapid Support Forces (RSF). Then, in April 2023, war broke out between the SAF and RSF, and the conflict has been deadly. As of late 2024, up to 150,000 people have been killed, 12 million have been displaced, and 22 m face severe hunger. The UN has described it as having spawned "the world's worst hunger crisis": country-wide famine – the worst in 40 years – was declared in August 2024 (*The Economist*, 2024).

It can be difficult to write about archaeology, conservation, and tourism during times like these. Yet heritage sites, museums, archives, and libraries have not

DOI: 10.4324/9781003491071-12

escaped the carnage (Hamid et al., 2023; Bradshaw, 2024). Does our continued work on these subjects suggest we care more about ruins, artefacts, and museums than about living people? Should we continue to work on heritage during wartime given such extreme human suffering? Is there a way in which archaeology, conservation, and tourism can continue to be relevant during conflict and even contribute to the well-being of conflict-affected communities?

Due to these events and the questions they foster, we have split our contribution into two sections. The first describes our conservation, management, community engagement, and tourism work until the onset of civil war in April 2023 (for excavations, see Emberling et al., 2025). This date is relevant because despite the challenges presented by the 2018–2019 revolution and 2021 coup, our project was able to function as normal; in contrast, the 2023 conflict has been the most destructive to human life and challenging to project implementation. The second section therefore presents a more critical reflection of how our project has pivoted to meet the difficulties of heritage research during this conflict. It discusses how the prevailing conditions have exposed project strengths and shortcomings, the latter requiring us to shift our focus and highlight the areas that require more financial and intellectual investment.

10.2 Jebel Barkal, Past and Present

Jebel Barkal is a sandstone outcrop 104 m in height located near the Nile River in northern Sudan. It is at the centre of an archaeological site whose ancient name was Napata and is located next to the modern town of Karima (Figures 10.1 and 10.2). Napata was an important settlement for more than 2000 years in antiquity. The earliest remains belong to a culture called Kush and date as far back as 1700 BCE. Napata was the southernmost outpost of the Egyptian empire from 1432 BCE until its collapse after 1100 BCE. The Egyptians built a temple complex in Napata focusing on their god Amun, but in a local form with a human body and a ram's head.

The Kushites returned to power after 800 BCE and made Napata their capital city, expanding temples, building palaces, and eventually building a series of royal pyramids for their kings and queens. After the collapse of Kush around 350 CE, the site was occupied by a medieval Christian community. Muslim groups would eventually begin to settle in the area around the 8th century CE, integrating with the local population. It is the descendants of these communities who now claim primary ownership of this region.

Today, this part of Sudan is inhabited primarily by Shaigiya, one of Sudan's many ethnic groups. The Shaigiya identify as Arab Muslims and ultimately trace their ancestry back to the Quraysh of Saudi Arabia. They are highly skilled sedentary date farmers, although many also make a living as professionals such as teachers, traders, writers, and politicians. Karima and surrounding villages are also home to people who identify as belonging to many other of Sudan's tribes, including semi-nomadic pastoralists, as well as much smaller groups of Christian Nuba and Nuer from the southern regions of Sudan and South Sudan, and Fellata,

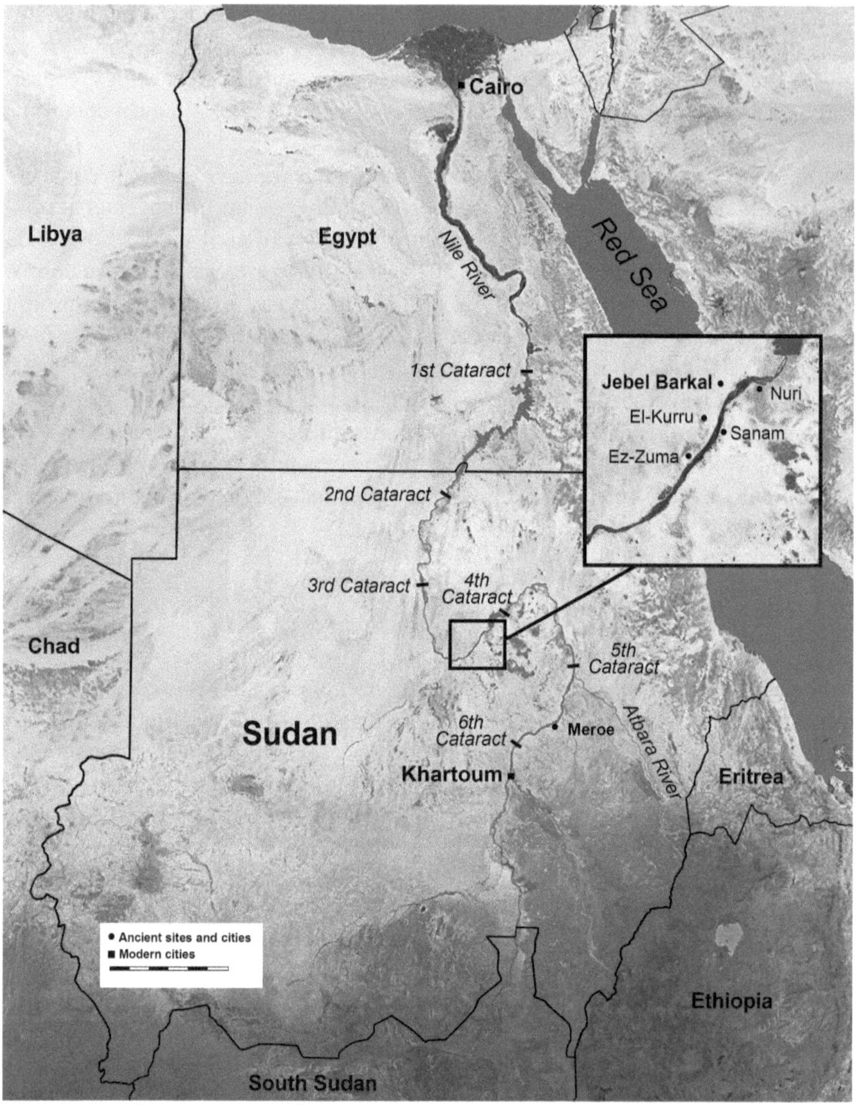

Figure 10.1 Map of Jebel Barkal and other sites mentioned in the text. Jebel Barkal Archaeological Project.

Note: Imagery and Map Data: Google Landsat/ Copernicus Data SIO, NOAA, U.S.Navy, NGA, GEBCO Maps GISrael IBCAO.

Figure 10.2 Jebel Barkal with the Temple of Amun, the Nile, and Karima in the background. © Kate Rose.

whose families migrated some centuries ago from West Africa. This diverse community is a vibrant mix of pre-Islamic, Islamic, Christian, and other pan-African traditions, many of which centre around Jebel Barkal itself.

10.3 Archaeological Research at Jebel Barkal

Jebel Barkal has been the focus of European and American archaeological interest since 1821, after the conquest of much of Sudan by the Egyptian army of Mohamed Ali Pasha. Later, in 1881, a Sudanese religious leader known as the Mahdi ('the Messiah') and his followers rebelled against Anglo-Egyptian colonisation and retook control of Sudan. They were defeated by the British in 1898. One of the colonisers' first steps was to conduct excavations of Sudan's archaeological sites, including Jebel Barkal. The largest-scale early investigations were those of George Reisner (1916–1920), who excavated pyramids, temples, and one of the palaces on behalf of Harvard University and the Museum of Fine Arts, Boston, which retains many of the finds from his work. As was standard practice at the time, conservation and site management were not part of activities at the site: 19th- and early 20th-century excavators tended to uncover architecture and artefacts and either removed them or left them in situ, unprotected.

As Jebel Barkal is a large and complex site, archaeological excavation there has been carried out by several international missions in the years since Reisner's work. An Italian team first resumed work at the site in the early 1970s and was followed later by an American team. Spanish and Sudanese teams have also excavated at the site.

Architectural conservation began later. There had been some mid-20th-century efforts at cement repairs in the temple of Amun and a smaller temple dedicated to Osiris-Dedwen (described further below), but these repairs do not appear to have been part of a coordinated conservation campaign, and the work was not documented. Beginning in the 2010s, a joint Italian–Sudanese conservation and restoration project was started in the rock-cut temple of the goddess Mut (Laurenti and El Malik, 2021), while the Italian and Spanish excavation teams have also done restoration work (Ciampini, 2022; Díaz de Cerio, in press).

As is clear from this brief history, the enduring obsession with architectural scale, both in Sudan and across the globe (e.g. Osborne, 2014; Harrison, 2013, pp. 42–67), means that research at Jebel Barkal has largely focused on its elite monumental structures: pyramids, temples, and palaces. While our project seeks to modify this approach by focussing on the ancient city and its non-elite architecture (see below), the appropriately designated 'outstanding universal importance' of these monuments led the National Corporation for Antiquities and Museums of Sudan (NCAM) to nominate Jebel Barkal and four nearby sites (El-Kurru, Sanam, Nuri, and Zuma) for inclusion on the UNESCO World Heritage List, becoming inscribed in 2003.

10.4 The Current Project: JBAP

In 2018, the current authors formed a new joint project between the University of Michigan and NCAM that we termed the Jebel Barkal Archaeological Project (JBAP: https://sites.lsa.umich.edu/jbap/). Our initial goals were relatively straightforward:

- To conduct excavations, expanding archaeological research beyond the monumental area of the site to investigate the ancient city itself and the lives of the non-elite.
- To begin conservation of the Amun temple area, which had been exposed by Reisner but never protected. This goal was an NCAM priority.

A further key aim was to work towards 'decolonising' archaeology and conservation fieldwork practices in two main ways. First, we embedded community engagement within each project component, as some of us had done in our project at the nearby site of El-Kurru (Emberling and Davis in press; https://sites.lsa.umich.edu/nubia/) and others at Sudan's second UNESCO World Heritage Site at Meroe (Humphris and Bradshaw, 2017; Abulgasim was also a director of this project). As mentioned above, community engagement is an increasingly important part of heritage research, and our experiences at El-Kurru and Meroe – both successes and failures – provided solid ground on which to launch engagements at Jebel Barkal.

Second, we ensured that each research and managerial position on the project was held jointly by a Sudanese scholar and an international scholar.[1] This pragmatic and, for Sudan, so far unique decision, demonstrates our conceptualisation of Sudanese and foreign colleagues as equal partners in planning and implementing the excavation and conservation work. This deliberate move was designed to increase all members' capacity for excavation, conservation, and site management at Jebel Barkal, to foster professional relationships and the exchange of knowledge between Sudanese and foreign team members, to improve project outcomes (including sustainability) and to equalise opportunities and benefits for team members.

Using this approach, we successfully applied for a grant from the US State Department's Ambassadors Fund for Cultural Preservation (AFCP) that would support community-engaged conservation at Jebel Barkal.

10.5 Conservation and Management

Early conservation and management efforts at the site began in earnest following the site's inscription, conducted primarily by UNESCO. State of Conservation Reports shared areas of concern for the site's conservation and management (UNESCO World Heritage Centre n.d.). These include the expansion of urban and agricultural areas in close proximity to the site; uncontrolled access to the site, especially by vehicles; impacts of changing climatic conditions and groundwater levels, most notably from the construction of a large dam on the Nile nearby at the 4th Cataract; the inherently fragile nature of the local sandstone from which Jebel Barkal's ancient monuments are constructed; and a lack of coordination among the many stakeholders invested in the site's long-term protection and management.

The site's first – and, to date, only – management plan, prepared by UNESCO but still in draft form, goes further in identifying conservation and management needs for the site (El Masri, 2007). It proposes site-specific solutions such as bollards to control vehicle access to the site, drainage for rainwater, and the removal of vegetation, trash, debris, and spoil heaps from historic excavations. The plan also calls out larger, organisational problems such as a lack of knowledge about international standards in conservation and, relatedly, a lack of formal academic curricula or training in conservation in Sudan. Similarly, it explores problems with financial, regulatory, and legal support for the site. Frustratingly, while the plan delves deeply into the many issues impacting the site's conservation, it provides few concrete steps for funding or implementing the suggested improvements. In places, the plan suggests that foreign missions working at the site could step in to fill the gaps in resources and coordination, but it also notes that foreign projects tend to operate independently, without a shared and comprehensive strategy for archaeological research, let alone conservation, site management, or tourism development.

Since the management plan was drafted in 2007, there has been progress. UNESCO has invested significantly in the protection of the site and has made repeated, if varied, attempts to coordinate state and local stakeholders in the site's protection and development. All foreign missions at the site now invest in conservation and restoration of specific monuments (see, e.g., Laurenti and El Malik, 2021; Ciampini, 2022; Díaz de Cerio, in press). These missions coordinate their work administratively with NCAM, including with input from NCAM's conservation team, and most have incorporated conservation training of NCAM personnel into their overall projects. Yet Sudan's fragile economic and political situation, and the geopolitical reality that funding for archaeological research and conservation is largely limited to the Global North, has meant that these projects have been staffed primarily by European/US-based professionals.

Our project has focused on conservation of two temples, with one of these, termed B700, having received the most attention thus far (Figure 10.3). Constructed between 650 and 620 BCE by King Atlanersa and his son King Senkamanisken, B700 is dedicated to Osiris-Dedwen, a syncretic god combining attributes of the Egyptian god of the underworld, Osiris, with an older Kushite god, Dedwen. It has two spacious, columned courtyards and a small chapel, all constructed from sandstone. Our conservation plans for this temple were set collaboratively by the team and have prioritised stabilising the temple's architectural remains and improving its legibility for visitors. In addition to conducting an initial needs assessment and extensive photographic documentation (including aerial photography and photo-modelling before, during, and after intervention), we have removed trash and vegetation, studied the original building materials, and replaced old, incompatible cement repairs with pigmented lime–sand mortar.

Figure 10.3 Temple of Osiris-Dedwen at Jebel Barkal after collaborative conservation work in 2023. © Sami Elamin.

Areas of undercut masonry were stabilised using red brick, lime mortar, and pigmented lime plaster. Fallen blocks were also reset where possible, using lime mortar. All significant fallen blocks that could not be reset were documented and numbered; uninscribed/undecorated blocks were buried within each courtyard to declutter the rooms and protect the blocks. In one courtyard, new sandstone flooring (easily distinguishable from the original by shape and colour) was installed in an area where original flooring had been removed during Reisner's excavations of 1916–1920, to make a discrete and more stable area for visitors to stand during tours. Our project has also funded installation of concrete bollards to control vehicular access to the site, removal of old spoil heaps left from Reisner's excavations, and programmes and activities for local community members.

In the course of our work, the joint Sudanese and American conservation team has shared a variety of information as we have worked to develop conservation plans. This knowledge exchange has ranged widely, from technical methods to supply chain logistics to aesthetic philosophy. We have discovered differences in our training and experience and, as a result, differences in how we understand various conservation methods. For example, infills perceived by Sudanese conservators as having a structural function were understood by the foreign conservators to be solely cosmetic. These differences necessitated ongoing discussions and sometimes required negotiations about desired outcomes.

10.6 Community Engagement

Community engagement is a professional practice that connects archaeological sites with local communities through methods such as education, training, and capacity-building to foster collaborative decision-making between these communities and the project team. As discussed above, it is a key tool used in the process of decolonising our fieldwork practices.

Sudanese archaeology has been slow to integrate community engagement into its normative practice. Several scholars – including members of this team – have written about why this has been and provided critical discussions of the projects that are now doing so (Humphris et al., 2021; Bradshaw and Emberling, 2022). Despite setbacks, community engagement has become increasingly mainstream in Sudanese archaeology since 2013, and we actively seek to embed it within our project.

Our community engagement work falls into three connecting categories: ethnography, mutual learning, and collaborative problem-solving. In line with best practice, we started with ethnography: meeting people for tea, introducing ourselves and the project, listening about people's daily lives and making observations, attending cultural events, and building relationships and trust to establish a strong basis for future seasons. We already had a robust community network as three of us (Abulgasim, Mohamed, and Elamin) are from the Karima region; the team also had well-grounded relationships with labourers working on our excavation and conservation projects (for their importance, see Mickel, 2021). We nevertheless wanted to meet people both inside and outside archaeology's obvious stakeholder groups and invested significant time 'hanging out' with everyone from tea ladies to footballers. Through asking members of local communities about their experiences with archaeologists and their projects, and in telling them about ours, it was clear that serious and systematic community engagement had never been conducted before, that this had caused some resentment in the local communities, and that the residents want to be consulted and involved in the project and its future goals.

At the same time, we initiated mutual learning. We began with school tours for those immediately adjacent to the site. We explained its Egyptian and Kushite history and – as our community engagement work has an explicit conservation aim – we also advised the students and teachers about site-appropriate behaviour including climbing monuments, trash dumping, making graffiti, and removing items from the site. These events offered opportunities for the school groups to inform us about their knowledge of the site (see also Abbas, 2020, and Abulgasim and Bradshaw, 2020), including rumours of tunnels underneath the mountain leading to other sites, and, stories and myths about jinn (benign supernatural beings that inhabit the ruins). They also told us about the myriad micro-experiences they have, such as passing the mountain each day on the way to the school or market, running up and down the mountain's sand dunes as part of their football resilience training, even having clandestine romantic moments, shielded by the high temple walls.

Using the basis of ethnography and mutual learning, we were able to identify both key issues of contention and key stakeholders with whom we could collaboratively problem-solve. We noted early on that community concerns often revolve around territorial issues: the encroachment of the site – and our/UNESCO's plans for the site – on land that the residents need to use. To take one example, Karima's municipal waste removal service is unreliable at best. For those living nearby, the site therefore provides a convenient location for trash disposal. In another example, the many people who own agricultural land adjacent to the site require their boundaries to be respected to fulfil the annual crop yields; any extension of buffer zones or excavation trenches has material effects on their income. In a final example, the site provides a quick cut-through for vehicles and donkey carts making their way from surrounding villages to Karima Hospital. In each case, unilaterally established and legally supported site boundaries can have detrimental effects upon residents who "rarely have access to reliable judicial channels through which they might contest potentially intrusive domestic and foreign projects – including those conducted by archaeologists" (Bradshaw and Emberling, 2022, p. 3).

These activities highlighted an important point (something we'd learned from El-Kurru and Meroe) that while much of the community is eager to connect with archaeologists and learn about the history of the site, their most pressing concerns are almost wholly detached from its 'authorised heritage discourse' (Smith and Waterton, 2009) – it's ancient Egyptian and Kushite history in which we, and our institutional funders, are most invested in. We therefore took collaborative action: we went with local residents to the municipal waste removal service to initiate negotiations about trash collection, we consulted with a landowner about our hopes of conducting geomorphology in sections across his palm grove and asked his permission, and we reorganised our plans for the restrictive bollards to include space for a small access road. We know from our relationships with community members that these are 'baby steps', but we have learned from our time at El-Kurru and Meroe that it's often better to walk before you run.

10.7 Our Approach to Sustainable Tourism

Our engagement with the surrounding communities as well as national and international authorities has naturally drawn us into discussions about sustainable tourism.

As we understand it, the aspiration behind sustainable tourism is for destinations to go beyond providing visitors with accommodation options, entertainment venues, and adventure activities. Rather, sustainable tourism dictates that the individuals involved in tourism and its aggregates make decisions that facilitate continuous, long-term tourism without detrimental effects to society and ecology. Even further, sustainable tourism encourages these individuals to make decisions that will actively alleviate poverty, promote gender equity, and support the environment (Giliberto and Labadi, 2022).

Sustainable tourism is therefore a concern of scientists and social scientists alike and its principles apply to everyone from hotel tycoons to tour operators. Heritage sites are a key context in which the principles of sustainable tourism might be applied, with archaeologists positioned as key implementers.

Residents in Sudan want tourism (Bradshaw, 2017; 2018). It is seen as long overdue, and a much-needed income stream. However, tourism in Sudan, even before the current war, was a *potential* phenomenon, not an existing one. Due to the perceived lack of security and the real lack of tourism infrastructure, international tourist numbers have always been low (Bradshaw, 2017). In 2020, a UNESCO consultant estimated that the Jebel Barkal region only received around 1,000 foreign visitors per year (Rebanks, 2020) compared, for example, with north-east Peru's 69,000 even as far back as 2003 (Brodie, 2010). A brief 64% increase in international tourism receipts in 2022 (UNWTO, 2022) is likely explained by relatives visiting Sudan after COVID-19 rather than an unprompted surge in tourism interest, and we certainly cannot quantify how this increase might have impacted Jebel Barkal. The situation is slightly different for domestic tourism, with Jebel Barkal having apparently received 3,500 Sudanese visitors in 2020 (Rebanks, 2020). Again, this number is still extremely low compared with the 1.2m domestic visitors to north-eastern Peruvian sites in 2003 (Brodie, 2010).

Despite this, there has been small-scale tourism development in Sudan, including in Karima. A luxurious rest house – known as the Italian Camp – has been built by an Italian company just outside the eastern line of the site's buffer zone, one of the local site guards has fitted up his house for guests, and, although it has been stalled for several years now, a hotel development has been initiated beyond the road on the northern line of the buffer zone. Overall, however, tourism enterprise remains firmly anchored in Khartoum: international visitors and wealthier domestic tourists who do come to Jebel Barkal arrive with tour operators based in the capital and, if not staying the night, will usually go to a handful of particular cafés in town for a meal before leaving.

Therefore, while tourism has figured heavily in Jebel Barkal's archaeological plans – including those of NCAM in its 2011 UNESCO Nomination File (Ahmed, 2002), its site management plan (El Masri, 2007), its Reactive Monitoring Mission (ICOMOS, 2019), and a more recent draft of a UNESCO Sustainable Tourism Strategy (Rebanks, 2020) – visitor numbers have been so sparse and inconsistent that it has been wholly unable to help alleviate poverty, promote gender equity, or support the environment.

Given that this was the context in which we began our project, we felt our mandate was to assist in pre-tourism preparation rather than to initiate and develop it. Moreover, we were aware that encouraging preparation for tourism development in this context could be detrimental: raising expectations without a robust ecosystem in place to provide long-term guarantees or reliable financial dividends. This healthy fear has been borne out in other locations; for example, Giliberto and Labadi's (2022) comparative analysis of multimillion-dollar heritage tourism development projects shows that such endeavours can fail even in viable and well-established destinations such as Morocco. Similarly, research in Turkey has also shown that tourism, especially what is known as short-stay tourism (when tourists visit for less than a day), can have negative impacts on communities (Orbaşli, 2013). Indeed, what we felt was possible was to prepare for a time at which there will be an upsurge in tourist numbers. Our project therefore has incremental goals to assist in the development of sustainable tourism, some of which naturally overlap with our conservation and community engagement plans. These include work described above, such as site stabilisation, making areas eligible for visitors, trash removal and relocation, and the elimination of spoil heaps; this is also one of the few areas we have been able to work on since the outbreak of war in 2023, as the next section will discuss.

10.8 Reflections: Conservation and Community Engagement in Wartime (Post-April 2023)

At the time of writing (late 2024), almost everything has changed for Sudan and, by very humble extension, our project. Karima has been deeply affected. Residents from here led the revolution in the north in 2018–2019, and many died or were injured during the SAF and RSF's brutal response to peaceful civilian protests. Martyrs from this time are lovingly depicted in the town's murals. Merowe, located across the river from Karima, was where the civil war began on 15 April 2023. Although the foci of war moved away shortly afterwards to Darfur and Khartoum, Karima has since received thousands of internally displaced people (IDPs) seeking refuge. IDPs live in cramped conditions within schools and camps; others live in the houses of Karima's residents, whose limited resources become stretched even more than usual. Inflation is insurmountable, unemployment and insecurity are pervasive, and fear for the future is endemic.

Amid this humanitarian crisis, we return to four key questions:

Do we care more about ruins, artefacts, and museums more than living people? Should we continue to work on heritage during wartime given such extreme human suffering?

The clear answer to this first question is no: our team does not prioritise ‘things’ over people, though we feel it is necessary to explicitly re-centre ourselves and our work on the basis of this ethical query. We argue that we should continue to work on heritage during wartime and that it is important and worthwhile. As Simpson (2009) has shown, people need heritage; it is vital for their well-being, and it can provide a sense of security during periods of uncertainty. It can also contribute significantly to the likelihood that IDPs will feel able to return to their homes after periods of displacement (Isakhan and Meskell, 2023) and contribute to restorative justice and rebuilding society in peacetime (Walters et al., 2018). This has been borne out in our context with the significant outpouring of concern generated by Sudanese residents who, with their smartphones, have been the key record makers of heritage destruction over the past months (Bradshaw, 2024). As trained and experienced archaeologists and conservators, heritage is also the area in which we are most able to help.

How can our project be helpful to both heritage and local communities during a period of conflict?

Over the past 18 months, it has become clear that the best way to maximise our impact during conflict is to continue our project with key modifications; unilateral withdrawal from a now well-established and well-funded project would be unnecessary and actively *un*helpful. It also provides us with new opportunities. For example, the influx of IDPs, many without work or school to attend, has meant the number of Sudanese visitors to the site has suddenly grown into some hundreds per week, particularly on Fridays and often with cars or trucks (Figure 10.4). The site provides a scenic and shady area for social gatherings; the tops of the pyramids are also thought to provide telephones with reliable internet connections. Of course, this increased footfall poses risks to the monuments, but it also provides opportunities for community engagement and to learn about how to manage large visitor numbers on site. This is not leisure tourism in a classic sense, yet it demands us to think about how to operate equitably during a crisis, whether via the provision of employment (more guards, local on-site tour guides), infrastructure (more shelters on site), and/or education that is both historical and preservation focused.

Figure 10.4 Large numbers of visitors to the site, August 2023. © Sami Elamin.

Has our project and its collaborative design successfully pivoted to meet the challenges of heritage research during wartime?

Collaboration with archaeologists on conserving an ancient Kushite site is obviously not a high priority for exhausted residents and traumatised IDPs. Carrying on 'as normal' is impossible even if day-to-day tasks still require completing. This is also true for NCAM members, many of whom have also been forced to leave their homes in Khartoum (where the NCAM main office is located), some transferring to Cairo. The Western members of our team have the luxury of remaining safely in our homes – a point worth reflecting on in itself. With our team thus dispersed and only Sami Elamin remaining at Jebel Barkal, we are struggling to protect the site, let alone effectively conserve it, in collaboration with a war-weary community. Indeed, our ability to continue has depended on Elamin's courageous and energetic presence. As the interim manager for archaeological sites in the entire Northern State of Sudan (which stretches from Karima some 400 km north to the Egyptian border and includes many of Sudan's largest sites) and also interim manager for the Jebel Barkal UNESCO site, he has an impossible job in the current situation. Among his projects has been moving the archaeological collection of the Merowe Museum across the Nile from Karima back to safer storage at Jebel Barkal (with funding from UNESCO). The government of Sudan has largely ceased functioning, so he and his remaining colleagues, including site guards, have no government salaries or budget to support any additional work.

However, since the outbreak of war, we have been able to repurpose some elements of our AFCP grant, thanks to flexibility by the director and staff of that programme, and we have sent funds to Sudan. With this funding, we have been able to support ten professionals: Elamin, five other archaeologists who have relocated to the Karima area, and four additional site guards. Together, they have coordinated community engagement and domestic tourism efforts including training student tour guides, organising site visits by 400 schoolchildren, some from displacement camps and some from villages in the region (Figure 10.5), as well as hosting several community groups (including a women's English language club) at the site. They have also continued two site management activities: screening and removing the massive piles of backdirt from Reisner's excavations in the early 20th century that obscure views and understanding of the site; the other continues a long-term (and expensive) project of erecting a ring of bollards around the site to restrict vehicle traffic, as had been envisioned in the 2007 UNESCO draft management plan.

Figure 10.5 School children and children from IDP camps showing their hieroglyph lessons and the Sudanese flag, October 2023. © Sami Elamin.

The answer so far is therefore 'yes' – our project has been able to meet the challenges of heritage research during wartime, by reshaping our priorities, redirecting funding, and training other Sudanese archaeologists on the ground.

What could we do better as we move forward? To what extent can we 'war-proof' our project?

Protracted and multi-stage warfare was not something our project expected; we were therefore admittedly ill prepared to address it. Some of us (Davis and Dafalla) have training in disaster preparedness and response, including knowledge of ways to protect cultural heritage sites during armed conflict. Based on this and now our lived experience, it seems impossible to completely war-proof a project. Yet, we were fortunate to have already agreed on a set of priorities – site protection and community engagement among them – that have continued to be equally if not more important during the war. We were and are also extremely fortunate to have Elamin present at the site. This point is key: to the extent that any 'war-proofing' is possible, it is the presence of experienced team members in-country and onsite that makes the difference. Despite the stress and risk this has entailed, it has helped keep the site safe while also providing much-needed income and a sense of purpose for those in Sudan. Continued work at the site has also kept the project team in close communication, strengthening our relationships.

The war has nevertheless highlighted systemic needs in site conservation and management. One of these is that more skilled craftsmen are needed, as well as in more regions. A single conservator or conservation architect can make great progress with site preservation, even during challenging times, but only if they have access to a skilled crew. Another is that some professions key to site conservation, like architects trained specifically in conservation and historic preservation, do not yet exist in Sudan. Further, Sudanese heritage workers need a broad range of conservation knowledge and skills – not only training in conservation of Nubian sandstone (which has often been the focus of conservation training programmes in Sudan), but also in preserving mudbrick, managing rain and groundwater on sites, investigating building materials, documenting and monitoring site and monument condition, and assessing and mitigating a wide range of risks.

Our project conservators' most important realisation, as we have tried to continue our project during the war, is that the ad hoc conservation of individual monuments and the conservation training provided by individual missions – even well-designed, collaborative missions like ours – will never be enough. Sudan needs formal academic curricula in archaeological conservation and site management, ideally as part of a dedicated degree programme. Archaeological conservation and site management will not be sustainable in Sudan until there is a steady supply of highly trained, in-country professionals, and during war and scarcity, the need for skilled Sudanese conservators is more apparent than ever.

To summarise, our work during the conflict has provided some support to displaced archaeologists and local labourers who had been working with us. It has also provided some meaning and connection to a shared heritage for people in the area whose lives have been disrupted. These contributions also serve to protect the monuments themselves so that they might survive the turbulence of the conflict.

Acknowledgements

We thank the NCAM, the Ambassador's Fund for Cultural Preservation of the US State Department, staff of the US Embassy in Khartoum, and the University of Michigan's College of Literature, Science and the Arts and Kelsey Museum of Archaeology.

Note

1 Elamin is currently the interim manager of the Jebel Barkal UNESCO World Heritage Site and has taken part in all areas of the JBAP project. For JBAP itself, El-Hassan and Emberling are co-directors; Dafalla and Davis are co-directors of conservation; and Abulgasim and Bradshaw are co-directors of community engagement.

References

Abbas, B.M.B. (2020). *Intangible Cultural Heritage around Jebel Elbarkal World Cultural Heritage Site*. Unpublished booklet.

Abulgasim, T. and Bradshaw, R. (2020). Understanding the Significance of Archaeological Sites in Sudan: A Case Study of the UNESCO World Heritage Site of El-Kurru, *Sudan Studies*, 62, pp. 77–90.

Ahmed, S.M. (2002). *Gebel Barkal and the sites of the Napatan Region*. UNESCO World Heritage Site Nomination File. Available at: https://whc.unesco.org/en/documents/168830.

Bradshaw, R. (2017). *Beyond Identity: The Socio-Economic Impacts of Archaeology upon a Non-Descendent Community in Sudan*. PhD diss., SOAS University of London.

Bradshaw, R. (2018). Fāida shenū? (What Is the Benefit?): A Framework for Evaluating the Economic Impacts of Archaeology, *Sudan and Nubia*, 22, pp. 188–197.

Bradshaw, R. (2024). *Sudan's War Devastates Cultural Heritage as Rival Generals Vie for Power*. Al-Monitor. Available at: www.al-monitor.com/originals/2024/06/sudans-war-devastates-cultural-heritage-rival-generals-vie-power

Bradshaw, R. and Emberling, G. (2022). 'Archaeology, Heritage and the Community in Sudan', in: S. Chirikure (ed.) *The Oxford Encyclopaedia of African Anthropology*. Oxford: Oxford University Press.

Brodie, N. (2010). 'Archaeological Looting and Economic Justice', in: P.M. Messenger and G.S. Smith (eds.) *Cultural Heritage Management, Policies and Issues in Global Perspective*. Gainesville: University Press of Florida, pp. 261–77.

Ciampini, E.M. (2022). *The Royal District of Natakamani at Napata, Restoration Report.* Available at: www.academia.edu/118246301/Restoration_Report_Jebel_Barkal_2022

Díaz de Cerio, M. (In press). Abasseya Project – Remarks on B2500 (Jebel Barkal). To appear in *Proceedings of the 15th International Conference for Nubian Studies.* Warsaw: University of Warsaw Press.

El-Masri, S. (2007). *Gebel Barkal and the Sites of the Napatan Region, Sudan, Draft Management Plan.* Unpublished. Version: v.1 – November 10, 2007.

Emberling, G. and Davis, S. (2024). 'Community Heritage and Archaeology at El-Kurru, Sudan: Amplifying Local Voices and Histories', in: D. Fisher-Livne and M. May-Curry (eds.) *The Routledge Companion to Publicly Engaged Humanities Scholarship.* Routledge, pp.141–157.

Emberling, G. et al. (2025). *Jebel Barkal 2018-2023: New Research on the Napatan and Meroitic Town.* Sudan & Nubia. pp. 73–98.

Giliberto, F. and Labadi, S. (2022). Harnessing Cultural Heritage for Sustainable Development: An Analysis of Three Internationally Funded Projects in MENA Countries, *International Journal of Heritage Studies*, 28(2), pp. 133–146.

González-Ruibal, A. (2017). 'The Virtues of Oblivion: Africa and the People Without Antiquarianism', in: B. Anderson and F. Rojas (eds.) *Antiquarianisms: Contact, Conflict, Comparison. Joukowsky Institute Publication 8.* Oxford: Oxbow, pp. 31–48.

Hamid, I., Yusef, M., Sabrin, I. and Fushiya T. (2023). *The First Report on Cultural Heritage Situations in Sudan. Sudan Heritage Protection Initiative (SHPI).* Available at: https://heritageforpeace.org/wp-content/uploads/2023/06/SHPI-final-report-27.06.2023.pdf

Harrison, R. (2013). *Heritage: Critical Approaches.* London: Routledge.

Humphris, J. and Bradshaw, R. (2017). Understanding the 'Community' before 'Community Archaeology': A Case Study from Sudan, *Journal of Community Archaeology and Heritage*, 4 (3), pp. 203–217.

Humphris, J., Bradshaw, R. and Emberling, G. (2021). 'Archaeological Practice in the 21st Century: Reflecting on Archaeologist-Community Relationships in Sudan's Nile Valley', in: G. Emberling and B.B. Williams (eds.) *The Oxford Handbook of Ancient Nubia.* New York: Oxford University Press, pp. 1127–1147.

ICOMOS. (2019). *Report of the Joint World Heritage/ICOMOS Reactive Monitoring Mission to The World Heritage Site of 'Gebel Barkal and the Sites of the Napatan Region',* Sudan. Available at: https://whc.unesco.org/en/list/1073/documents/

Isakhan, B. and Meskell, L. (2023). *Mosul Faced Mass Heritage Destruction by the Islamic State. We Asked Residents What They Thought about Rebuilding.* The Conversation. Available at: https://theconversation.com/mosul-faced-mass-heritage-destruction-by-the-islamic-state-we-asked-residents-what-they-thought-about-rebuilding-207725

Lane, P. (2017). 'Envisioning a Different Notion of "Indigenous Archaeology" from the Perspective of Sub-Saharan Africa', in: C. Hillerdal, A. Karlström, and C.-G. Ojala (eds.) *Archaeologies of 'Us' and 'Them': Debating History, Heritage and Indigeneity.* London: Routledge, pp. 113–26.

Laurenti, M.C. and Osman El Malik, E.M. (2021). *Restoring the Colours of the Goddess: The Conservation Project of the Temple of Mut at Jebel Barkal.* Rome: Gangemi.

Mickel, A. (2021). *Why Those Who Shovel Are Silent: A History of Local Archaeological Knowledge and Labor.* Louisville: University Press of Colorado.

Ndoro, W., Chirikure, S. and Deacon, J. (2018). *Managing Heritage in Africa: Who Cares?* Abingdon: Routledge.

Orbaşli, A. (2013). Archaeological Site Management and Local Development, *Conservation and Management of Archaeological Sites*, 15(3–4), pp. 237–53.

Osborne, J. (2014). *Approaching Monumentality in Archaeology. IEMA Proceedings 3.* Albany: State University of New York Press.
Rebanks, J. (2020). *Sustainable Tourism Strategy for Gebel Barkal and the Sites of the Napatan Region*. Unpublished.
Schmidt, P.R. and Pikirayi, I. (2016). *Community Archaeology and Heritage in Africa: Decolonizing Practice*. Abingdon: Routledge.
Simpson, M. (2009). Museums and Restorative Justice: Heritage, Repatriation and Cultural Education, *Museum International,* 61(1–2), pp. 121–129.
Smith, L. and Waterton, E. (2009). *Heritage, Communities and Archaeology*. London: Duckworth.
The Economist (2024). Anarchy in Sudan Has Spawned the World's Worst Famine in 40 Years. Available at: www.economist.com/briefing/2024/08/29/anarchy-in-sudan-has-spawned-the-worlds-worst-famine-in-40-years.
UNESCO World Heritage Centre. n.d. Gebel Barkal and the Sites of the Napatan Region. Available at: https://whc.unesco.org/en/list/1073/.
UNWTO. International Tourism Back to 60% of Pre-Pandemic Levels in January-July 2022. Available at: www.unwto.org/news/international-tourism-back-to-60-of-pre-pandemic-levels-in-january-july-2022
Walters, D., Laven, D. and Davis, P. (2018). *Heritage and Peacebuilding*. Cambridge: Cambridge University Press.

11 Co-Production Networks for Community Heritage in Tanzania (CONCH)

Paul Lane, Elinaza Mjema, Noel Lwoga and Stephanie Wynne-Jones

11.1 Introduction

The coastal zones of Tanzania and its wider region have a rich tradition of vernacular architecture, centred on 'stone towns' that were constructed along the Indian Ocean coast from the 11th century onwards (Wynne-Jones and LaViolette, 2018; Sadik et al., 2023). Many still thrive today, and the inhabitants are rightly proud of their heritage, which includes UNESCO World Heritage sites at Kilwa Kisiwani, Songo Mnara, Zanzibar and Lamu. All are still home to Swahili groups, and the coast retains a distinctive culture based on urban living, Islam, coastal economies and the Swahili language. Yet the Swahili coast is also economically underdeveloped and marginalised within the nation-states of Kenya and Tanzania (Kabyemela et al., 2024).

Tanzania is one of the least developed countries on the Organisation for Economic Co-operation/Development Assistance Committee (OEC/DAC) list (OECD, 2024). Lack of employment and poverty are widespread. Rapid change caused by urbanisation and modernisation is leading to loss of both tangible and intangible heritage (Flores de León et al., 2020), as first highlighted some 15 years ago in the Tanzania government's 2008 Cultural Heritage Policy (URT, 2008), which stressed inadequate public awareness and the absence of cultural heritage in the school curriculum as being particular shortcomings. That policy document also noted the weak participation of local authorities and communities in conservation efforts at the time, the need for clear structures at village and district level, and the importance of community participation.

Tourism features prominently in Tanzania's economy. It is the country's second largest foreign exchange earner, based primarily on natural landscapes and wildlife tourism, supplemented by beach holidays (Wamboye et al., 2020). Tourists are often unaware of the region's rich and culturally diverse past, as is common in many other African nations (Chirikure et al., 2021), while a lack of education among residents makes it difficult for cultural heritage sites to be presented to their full potential. Despite a growing recognition of the need to engage local communities in heritage efforts in Africa (Chirikure and Pwiti, 2008; Lwoga, 2017a), there have been few initiatives to mobilise value for local stakeholders through built

DOI: 10.4324/9781003491071-13

heritage, with a particular neglect of the potential heritage value of the built architecture dating to the 19th to early 20th century CE.

Such an oversight may be due in part to the majority of surviving standing buildings from that era having associations with either foreign colonial rulers, initially Omani Arabs (Pradines, 2004), and subsequently German and British authorities (Rhodes, 2016), or with other foreign groups, such as merchants from western India, with some buildings also having intimate connections with the Indian Ocean slave trade (Mapunda, 2008). Moreover, while the national authorities for both mainland Tanzania and Zanzibar have made efforts to protect and even renovate a selection of these buildings, notably in Zanzibar Town (Khalfan and Ogura, 2012; Heathcott, 2013), Bagamoyo and Kilwa Kivinje (Ichumbaki, 2016), urban expansion and renewal, often unmonitored within the planning process, have led to either demolition or significant decay of these buildings over the past 60 to 70 years, especially in Dar es Salaam (Kigadye, 2014; Swai, 2023). This process has been exacerbated by the benign neglect of former German and British colonial buildings used by local government administrations since Independence.

In part due to legislative uncertainty, but also owing to limited resources from the state for their conservation and repair, much of the built heritage that survives is in a poor state. With post-Independence urban development and the expansion in some coastal settings of tourist hotels and other infrastructure, the integrity of many historic streetscapes and urban landscapes has been disrupted, such that at the time the 'Co-Production Networks for Community Heritage in Tanzania' (CONCH) project was being designed, some of the best surviving streetscapes of late 19th- and early 20th-century architecture in the country were located in small towns, such as Pangani and Kilwa Kivinje, poorly connected to the national road network and in long-term economic decline. An important motivation for the CONCH project, therefore, was to explore ways of demonstrating how the built environment can be a crucial resource for sustainable socio-economic development, heritage tourism and urban renewal.

11.2 The Co-Production Networks for Community Heritage in Tanzania (CONCH) Project

CONCH was first conceived as a research network with initial support from the UK's Arts and Humanities Research Council (AHRC). Subsequently, additional funding was provided via the University of Cambridge's allocation from Research England's Global Challenges Research Fund Quality-Related (GCRF-QR) funding, Cambridge University's internal Arts and Humanities Impact Fund (AHIF) and a University of York Impact Accelerator Fund (YIAF) award. The Cambridge funding was secured specifically to support the preparation of educational resources explicitly requested by local teachers in Pangani following initial work there by CONCH, and in collaboration with ArchaeoLink, a UK-based archaeological trust, from 2019 to 2021. YIAF funds supported a Swahili language book about Kilwa, also a response to specific requests (see below). The AHRC-supported phase ran for two years from mid-2017 to mid-2019 and involved work with diverse stakeholders in the coastal towns of Pangani and across the Kilwa archipelago, northern and southern Tanzania, respectively (Figure 11.1).

During this phase, CONCH worked to create a structure for engaging researchers, heritage professionals, local businesses and community groups, focusing on the built heritage of the Tanzanian coast and its urban settings. CONCH's objectives were to enhance recording, documentation and appreciation of the built heritage of these towns, including that of colonial origin, and identify how these assets might be used for economic development and improved social well-being. At the same time, the project was concerned with demonstrating the value of integrating knowledge co-production into designing heritage protection measures.

As part of its activities, CONCH sought to make the heritage practices of community members more visible and to explore the use of heritage in response to current environmental and societal challenges. By building better partnerships between researchers, NGOs, community groups, government officials and other heritage stakeholders, the network aimed to improve the conservation of historic buildings and urban landscapes in both Pangani and Kilwa. In Pangani, CONCH was facilitated by Uzikwasa, a local NGO, which under the direction of its then Chief Executive Officer, Dr Vera Pieroth, had already established a Heritage Centre (Pangarithi) in the town and had secured funding from the Tanzania US Ambassador's Fund and other sources. Uzikwasa had commissioned a detailed assessment of Pangani's heritage resources, architectural surveys of the standing historical buildings, and the installation of interpretation panels adjacent to key historical buildings (Moon and Blanchard, 2009). In Kilwa, CONCH worked with partners across the Antiquities infrastructure, particularly Mercy Mbogelah, the site manager for Kilwa, but also the Ruins Committee chairpeople for Kilwa Kisiwani and Songo Mnara.

The CONCH project began with an online kick-off meeting where key project participants from Europe and Tanzania met and agreed on work plans and project protocols. This was followed by an in-person two-day workshop at the University of Dar es Salaam (UDSM) in the Department of Archaeology and Heritage Studies in mid-2018. The workshop involved academic partners, collaborators from international heritage organisations (UNESCO), local heritage organisations (Uzikwasa, DARCH), national bodies (Antiquities Department of Tanzania) and community leaders from Pangani. Topics covered in the workshop included: a review of the heritage assets of Pangani and Kilwa Kivinje and especially their historic buildings; the principles and methods of buildings archaeology; the potential value of geophysical survey for detecting traces of earlier urban development; the impact of past oceanic catastrophes and climate changes (Mjema, 2018); the emerging field of digital heritage and its potential for enhancing public understanding of heritage assets; and the scope for developing educational resources for primary and secondary school teachers and students.

A key outcome from this workshop was recognition of the importance of involving all stakeholders, especially community heritage activists, in matters related to coastal heritage and formation of network activities. The workshop also served to create a structure for engagement of local communities in cultural

Figure 11.1 Location of the CONCH Project study sites in coastal Tanzania. © Stephanie Wynne-Jones.

heritage sectors in coastal Tanzania. It helped stakeholders to understand how they could participate in the CONCH project and to identify some of the metrics that might be used for measuring the progress of the project.

11.3 Professional Training Field Schools in Pangani

Several training field schools took place at Pangani to familiarise local scholars and students with techniques of mapping and documenting built spaces. The project team ensured that local community groups including schools were incorporated in the process and were encouraged to participate in finding out about their region's past. Local business owners, such as hoteliers and small craft producers, were also incorporated, allowing them to build cultural heritage tourism into their activities.

11.3.1 Buildings Archaeology

A professional building archaeologist on the team, Dr Dav Smith (University of York) provided training in Pangani on how to record, analyse, and interpret standing historic buildings. Buildings archaeology as a distinct field is not taught in Tanzania and was unfamiliar to most heritage professionals at the time of the project. To generate awareness of the field, instruction focused on how to look at standing buildings archaeologically and how to create accurate measured drawings, such as plans, elevations, and cross sections. The practical exercises in the field also involved learning about the region's architectural traditions that blend Swahili practices with those imported from India, Oman, Germany and Britain. The final exercise involved the creation of 3D models of standing buildings, including earthen structures, using 'Structure from Motion', or SfM, photogrammetry software. The building recording and reconstruction activities brought together different conceptual threads that touched on the influence of authorised heritage discourse on determining what constitutes a 'historical building', local knowledge of buildings, and archaeological perspectives.

11.3.2 Geophysical Survey

Geophysical surveys of open spaces in Pangani were carried out in July 2018 in different areas of the town including around the Boma building, Uhuru Ground and Pangani's football ground. To enhance local skills, this element of the project involved training of Uzikwasa's heritage officer (Mr Severine Kizito) and the local Tanzania Antiquities Department officer (Mr Ali Mbarouk) based at Tongoni Ruins, south of Tanga. The surveys were led by Professor Kate Welham (Bournemouth University) assisted by an MA student, Mr Hayden Scott-Pratt, and involved both magnetometer and fluxgate gradiometer surveys. Topography of the mapping of the survey areas was overseen by Mr Nicholas Pearson (On-Site Archaeology, York). The survey of Pangani football ground, on the edge of town, failed to detect any anomalies, but surveys of some of the other areas suggested buried archaeological horizons in those locations.

11.3.3 Test Excavations and Watching Briefs

Pangani has often been considered as having been founded in the late 18th or early 19th century in response to the growth of the caravan trade, especially in elephant ivory and enslaved individuals, replacing an earlier Swahili stone town (Muhembo) situated on the bluff to the north. During the 19th century, the town grew in prominence and served as an important transshipment hub as well as a later centre of revolt against German overrule (Glassman, 1995). Previous archaeological surveys and excavations (Gramly, 1981; Mjema, 2015; Rhodes, 2010; Walz, 2010) around Pangani have highlighted that the area has a long history of human settlement and trading activities, while test pits excavated at various points within Pangani by Rhodes (2010) and Walz (2010) and under the auspices of the Historical Ecologies of East African Landscapes (HEEAL) project (Lane, 2010) with support from On-Site Archaeology (On-Site Archaeology, 2010, 2013) indicate the presence of settlement traces predating the 19th century. To further investigate the extent of pre-19th-century deposits at Pangani, a small test excavation directed by Mr Graham Bruce (On-Site Archaeology) was undertaken by the CONCH team immediately adjacent to the Old Boma. This is a building that had once served as the centre for German and British administrations and is now used to house local government offices.

This excavation aimed to establish when the Old Boma had been built and whether it had replaced any pre-existing structures. Given its central location adjacent to the public open space of the Uhuru Ground, it was also an ideal place to run an excavation accessible to members of the public and school visits, while also highlighting the work of the CONCH project to local government officers. The excavation also provided the opportunity to train Uzikwasa's heritage officer and the local Tanzania Antiquities officer in the principles of single context recording. The excavation was a great success and attracted a lot of public attention, especially from schoolchildren who visited both as part of organised groups and on their own accord. The excavation uncovered the remains of several wall foundations associated with cultural artefacts including Omani-era pottery, German metal objects and bottle glass. These were catalogued and photographed, shown to local residents, and included in a small display at the Pangarithi Heritage Centre.

On arrival in Pangani in 2018, the CONCH team noted that wall foundation trenches had been excavated for a new building being erected in the grounds of Old Bohra Mosque, now a madrassa for the Bilal Muslim Mission of Tanzania, just north of Pangarithi on Lumumba Street. After receiving permission from the owners, this provided an excellent opportunity to demonstrate the value to both the local Tanzania Antiquities officer and Uzikwasa heritage officer of undertaking watching briefs wherever the subsurface is disturbed by construction work (Figure 11.2). This involved inspecting the trenches that had been excavated and documenting the location and nature of visible archaeological remains and completing a short watching brief report. As anticipated, wall foundations were recorded in several trenches and a great deal of pottery was recovered from the

Figure 11.2 Watching brief inspections at Old Bohra Mosque. © Paul Lane.

upcast, a small sample of which is now at the Heritage Centre for public display. Since the end of the CONCH project, the Uzikwasa heritage officer has continued to monitor building work in Pangani and elsewhere in the district.

11.3.4 Public Archaeology

One of the goals of the CONCH project was to ensure that university students, and primary and secondary school pupils in Tanzania are familiarised with the types and importance of the built heritage that surround them, specifically that found in Pangani and at Kilwa on the Tanzanian coast. Students who undertake a BA degree in Archaeology and Heritage Studies at the UDSM are required to spend a total of 112 days in the field before graduation. The UDSM field schools help them to have longer and direct interaction with the cultural heritage resources found in different parts of the country. The benefits and knowledge that UDSM students acquire during their field schools typically are not shared with their juniors in primary or secondary schools.

To ensure that the CONCH project benefited students at different levels, students involved in UDSM archaeological field training around Pangani Bay, which in 2018 focused on Kikokwe Fort, were given responsibility for guiding students and teachers from a nearby primary school at Kikokwe during a day-long visit. The primary school students and teachers spent the entire day with the UDSM field team walking around the site, listening to and helping team members and telling the UDSM university students what they knew about the area.

The UDSM students provided a brief orientation about the history of the fort and the aims of the project and spoke about the importance of preserving Pangani's built heritage for the benefits of history and tourism. The UDSM students also explained how they documented the site and explained why the project had decided to excavate outside and inside the fort. They pointed out that these excavations aimed to reveal something about the communities that lived in the area before the fort was built. The direct interaction and communication between the UDSM archaeology students and their juniors at the fieldwork site hopefully brought benefits to them both, introducing the primary school students to archaeology and heritage, and even think of these as promising disciplines for future study, while also helping promote interest in the conservation of Pangani's tangible heritage.

11.3.5 ArchaeoLink School Project

Ms Patricia Hart (ArchaeoLink) worked with CONCH project researchers and Pangani community members to develop educational programmes that could showcase Pangani's rich built cultural heritage history while also advancing basic skills and communicating principles of sustainable development. Practical activities were developed in consultation with teachers at seven local schools to help build skills in observation and general knowledge. These took the form of 'heritage walkabouts' involving CONCH team members and the Pangarithi Heritage Office in groups of 10–20 pupils and their teachers (Figure 11.3). Stops were made at selected buildings and students were asked to sketch what they saw, while being briefly told about the history of the building. The aims were to facilitate knowledge exchange and encourage an appreciation of built heritage, while also having fun.

The students were encouraged to look at different details and to think about the building materials and methods of building. For example, students were asked to look at the coral stone, mangrove poles, high ceilings and thick walls and were asked questions about where they thought these building materials were obtained and how the design of the building kept buildings cool despite the absence of fans or electricity. The trips ended at Pangarithi where the children looked at the photograph of Pangani's 1900 waterfront and identified buildings that they had visited. In all, five walkabouts were attended by a total of 44 pupils from all seven schools, some coming as many as four times over the weekend. Some 100 sketches were produced, and several were exhibited at Pangarithi. The students were allowed to keep their sketchbooks and pencils in the hope they will continue to sketch and appreciate their heritage.

The activities generated considerable interest among the teaching staff at the participating schools, prompting requests for educational materials. Once additional AHIF and GCRF-QR funding was acquired, ArchaeoLink was commissioned to produce a suite of materials comprising 40 comprehension exercises, in English, on themes covering the history and architectural heritage buildings of Pangani, 20 activity sheets, in English and KiSwahili (for use both in class and outside the classroom), six wall posters (on local archaeology and history), a map quiz of heritage buildings, and an interactive board game modelled on snakes and ladders

Figure 11.3 School walkabout to observe Pangani's historical buildings. © Paul Lane.

(Figure 11.4) featuring Pangani's diverse tangible heritage, and comprehensive teachers' notes in Kiswahili and English on all of the materials and on asking stimulating questions. These materials were designed according to Bloom's revised taxonomy of higher order thinking skills in a manner that provided students of all abilities the opportunity to think at progressively higher levels with each activity (Ormell, 1974–1975).

Complete sets of these materials were supplied to all seven participating schools with an additional set supplied to Pangarithi. To familiarise teachers and others with the materials and how they might be used across the curriculum to support critical thinking, a series of workshops and training events were held in Pangani in July 2021 (after delays due to the COVID-19 pandemic), involving 77 teachers (mostly for history, English and civics; 44 primary and 33 secondary teachers) from local schools, the Uzikwasa Heritage Centre officer, the Go Pangani representative and seven graduate students from the Faculty of Education, University of Dar es Salaam. Meetings were also held with the heads of the various schools, Uzikwasa and several local government officers including the District Commissioner for Tanga, the Executive Director, the Director of Primary Education, the Director and Assistant Director of Secondary Education and District Cultural Officer for Pangani District. All gave their support to the project, and the Director of Primary

Figure 11.4 A wall poster providing a quiz on identifying historical buildings in Pangani, produced for local schools by ArchaeoLink. © Patricia Hart, ArchaeoLink.

Education encouraged the project to expand the number of schools it collaborates with. By providing training and teaching resources, the project introduced new ways of teaching in primary and secondary contexts and helped expand their understanding of what a 'competence-based curriculum' entails. The project has subsequently offered this same training and practical experience to graduate students in Education at the University of Dar es Salaam. These will either qualify as teachers or be employed in other parts of the educational sector, and in providing this, the project has sown the seeds for more widespread adoption of child-centred learning.

11.3.6 Local Radio

To further promote awareness of the project and Pangani's heritage, the CONCH project team had a heritage walkabout for a radio show hosted by Pangani FM, a local radio station established by Uzikwasa. This was led by Dr Sara Perry (then at the University of York) and her Heritage MA students from York, with other collaborators, including Pangani FM presenters. To create the radio show in two days, four buildings were selected. These were the Pangarithi building, which at the time was used as a heritage centre and also known as Rehema Abdul (or Bwanga) House, but was originally a two-storey mixed residential and commercial property for a prominent merchant family in the 19th century; Al-Kasaby CCM House, now derelict and partially collapsed at the time of the project, but reputedly associated with a former prominent slave trader; the Pangani Industries building, also derelict, but formerly an imposing residential stone building facing the waterfront; and the former Regal Cinema built in 1950 and operational until the 1980s, but now partially derelict and used as a vehicle repair shop. Each member of the team chose one building to research using a variety of documents and images. From this, a brief narrative of each building's history was created to introduce it along with a set of questions the radio presenters could use to prompt responses from community members joining the show, who included Pangarithi's supervisor, a local resident and the village chairperson. Following this, the project's buildings archaeologist provided some comments on each of the buildings and their heritage value.

11.3.7 Digital Heritage

The CONCH heritage team set up a digital Twine app to help reveal the hidden stories of artefacts held in Pangarithi, in response to Uzikwasa expressing a desire for the creation of interactive digital resources. The Twine app was created collaboratively with key partners. Together they identified several objectives in line with Uzikwasa's aims of improving the quality of life around Pangani. A design process was subsequently agreed upon wherein higher-level ideals were identified from which specific components of the proposed experience were developed. This enabled consideration of the vision and objectives discussed in the original meeting with Uzikwasa against the wide variety of options available for developing a specific interactive experience. The group decided on six themes that resonate with positive emotions should provide the conceptual framework. These were hope – for now and for the future; respect – for people of the past; curiosity – to learn more about the people of Pangani; desire – for the objects; excitement – about Pangani's many and complex histories; and empathy – feeling a connection to and being able to take the perspective of someone from the past. By stirring these emotions and others, the aim was to create experiences that would lead visitors to encourage others to visit and to make return visits in the future and ideally raise the visibility of – and take part in – other activities offered by Uzikwasa and Pangarithi.

The team then selected objects on display at Pangarithi. Some of these artefacts were quite expressive of individual lives and activities – e.g. a key, a button, broken plates – but their stories were not yet being told to visitors. Accordingly, the team designed an interactive experience that was intended to be eye-catching and playful, using colours associated with water such as blue and green due to Pangani's geographical location by the sea and along a river, and which could be viewed and shared by multiple people simultaneously.

The prototype used simple stories with choices of multiple endings as a way of exploring the possible biographies of different objects on display in Pangarithi. Woven into these stories were larger decisions and questions about visitors' values and interests in the present – which connected to the different paths that could be followed within the narrative. These decision-making tracks were recognised by Uzikwasa for having the potential to encourage deeper reflection and action and offering scope for developing leadership skills among its members and community stakeholders.

11.4 CONCH Activities at Kilwa

Kilwa District in southern Tanzania offers a very different setting to Pangani in terms of archaeological and heritage resources. It is home to a World Heritage site made up of the twin towns of Kilwa Kisiwani and Songo Mnara, which together contain the architectural ruins of some of the grandest examples of Swahili coral architecture of the 11th to 18th centuries (Garlake, 1966; Wynne-Jones, 2018). The sites, occupying two small islands, have drawn significant scholarly attention for their rich archaeological remains that have defined our understanding of coastal society (Chittick, 1974; Horton et al., 2022). Since their joint World Heritage designation in 1981, they have also been the focus of numerous heritage interventions. Kilwa and Songo Mnara were placed on the UNESCO endangered list in 2004, which prompted a series of internationally funded projects designed to save and conserve the built heritage of the archipelago; their success is attested by the sites' removal from the list in 2014 (Ichumbaki, 2016). These projects have largely worked to stabilise the buildings at Kilwa Kisiwani and Songo Mnara; despite changing funding regimes they have worked locally via the Antiquities infrastructure. Kilwa has a dedicated Antiquities office, a regional branch of the Antiquities Division of the Ministry of Natural Resources and Tourism. Conservation work at the World Heritage site is managed by staff from Antiquities, who work through a local team of labourers derived from the surrounding village; a local staff member has been trained to direct additional buildings conservation work and manages local teams when required.

During these various campaigns, work on the islands has also resulted in the creation of 'ruins committees' at both Kilwa and Songo Mnara; they act as a community voice on local interventions and help with the organisation of labour as and where required (Wynne-Jones and Fleisher, 2015). In recent years, the focus of work at the sites has turned towards promoting cultural heritage tourism, especially since 2016, when the Kilwa Antiquities branch was put under the authority of the Tanzanian Wildlife Management Authority (TAWA) as part of a restructuring of the Ministry that allocated many heritage sites to different government authorities in a bid to invigorate tourism more generally.

Members of the CONCH team travelled to Kilwa in 2019, with a similar set of aims to those from Pangani, adapted to fit the very different local context. At Kilwa, significant heritage infrastructure exists and there is a history of work addressing community engagement in the region. Most notably, national policy has been designed to incorporate local stakeholders at all levels. The *Tanzania Development Vision 2025* is interpreted at Kilwa as allowing local communities to profit from their involvement in the cultural heritage sites (TAWA 2023, 15). The Integrated Management Plan for the sites lists dozens of local stakeholders, including fishing communities, teachers and religious leaders, alongside the official government agencies (TAWA 2023, 13). This can also be seen in previous heritage interventions at the site, where communities and community benefits have been explicitly foregrounded in conservation campaigns (Ichumbaki and Mapunda, 2017).

The team brought specific skills in the digital recording and presentation of archaeological heritage via Dr Sara Perry and Dr Dav Smith (University of York). The team was led by Stephanie Wynne-Jones who has a history of archaeological work with communities around the Kilwa region since 2001 with a network of connections. The 2019 trip had a series of objectives building from these core skills.

11.4.1 Training and Capacity Building Around the Monitoring of Built Heritage

Staff at Kilwa Antiquities office had identified the need for increased monitoring of standing structures, to prevent decay and collapse of buildings. In particular, the staff were concerned to explore technologies that could be used at Kilwa Kivinje. This town, 12 kilometres north of the main cluster of Kilwa sites, flourished during the 18th and 19th centuries as a port initially built by Omani colonial authorities and then taken over by the Germans. Various coral-built structures relating to this history, notably the customs house, are still extant but in a poor state of repair (Figure 11.5), and the town has not seen the same conservation investment as the precolonial sites.

Kilwa Kivinje therefore served as the setting for training and testing of methods of conservation survey. Using a standardised recording form, the team conducted a visual survey of the former customs house known as the *boma*, recording the nature and extent of cracking, areas of vegetation ingress and areas where coral was deteriorating or damaged. This was achieved using paper recording sheets with

Figure 11.5 The customs house, Kilwa Kivinje. © Stephanie Wynne-Jones.

standard headings. Although simple, this system is easily repeated and can be done every 6 months to monitor deterioration.

To complement this paper record, Antiquities staff were trained in using photogrammetry to record standing structures. We recorded several sections of the building completely, using repeated photographs which were later processed using Structure from Motion (SfM) software to produce accurate records of the walls. Digital cameras were provided to Antiquities staff to continue this work after the team's departure.

11.4.2 Digital Tools for Presenting Kilwa's Heritage

The CONCH team did not set out to create a significant digital toolkit for Kilwa's stakeholders, that being unfeasible within the parameters of the network grant and our time in the field. Instead, our aim was to understand what tools already existed and to work with stakeholders to identify possible future toolkits that would fit their needs. We also wanted to explore the ways that previous digital provisions were being used. In fact, the Kilwa region has seen many interventions using digital technologies. The ruins of Kilwa Kisiwani and Songo Mnara have been laser scanned multiple times, resulting in a set of 3D images available on Sketchfab (CyArk, 2024; St Andrews University, 2021; Zamani Project, n.d). In addition, capacity development work by an EU team produced a Facebook page with images and text about the sites of Kilwa region; these are reproduced in posters in the Antiquities office.

Interviews were conducted with several groups, including the Antiquities staff, independent tour guides operating as part of 'Kilwa Island Tours', hotel owners and local communities. All were keen to use digital technologies, especially social media to advertise the sites and give information for tourists about how to reach them. The Facebook page created by the EU project cannot now be used for this because the login and password were not shared locally and are unrecoverable. This is an example of some of the challenges with digital technology projects, which are not sustainable as the tools for ongoing management are not accessible locally. The CONCH team therefore focused on social media training, working with Antiquities and with Kilwa Island Tours to create a series of Instagram posts and plan for future updates that way.

Communities on the islands of Kilwa Kisiwani and Songo Mnara consistently expressed a desire for better communications using digital technology. For them, this was not about communicating with global tourist markets but to empower island communities to access economic opportunities coming from heritage and tourism. Across both islands, but especially Songo Mnara which is in less frequent contact with the mainland, the inhabitants asked for advance notice of tourist visits, enabling them to cook food, or bring out craft objects for sale. They requested digital technologies that could achieve this. This challenge has now been taken up by a research student at the University of York, who is working on a SMS-based system for communications across the region (Bettinson, 2024).

11.4.3 Community Voices in Heritage Resources

Interviews with island residents, and with guides from Kilwa Island Tours, also highlighted a lack of information available locally about the heritage sites. The inhabitants of the islands felt disconnected from their own heritage, with no information in the Swahili language about the sites themselves or the history of the area. They also felt disconnected from tourists, unable to communicate community concerns during tourist visits.

The CONCH team conducted focus groups with island residents. We interviewed separate groups of women and men from each island, to capture a range of perspectives. We also spoke with village groups who were working with Antiquities, as site custodians. During these interviews, we sought to establish community perspectives on the sites, asking questions such as 'When your family visits, what do you most want to show them?' and 'What is your favourite building and why?' to stimulate discussion. The answers to these questions were incorporated into site information leaflets created by the CONCH team before leaving Kilwa.

These are deliberately low-tech leaflets, printed in black and white on a sheet of A4, which can easily be photocopied in Kilwa Masoko. The leaflets are available in both English and Swahili. They include some of the community opinions on what is important about the site. They also include a series of answers to another of our questions: 'What would you like to tell tourists?' The answers tend to revolve around two concerns. First, island communities requested that tourists adopt modest dress and respect local customs of greeting and hospitality. Second, they asked us to make tourists aware that they could buy drinks, snacks and crafts locally, rather than bringing them to the site. All of these are incorporated into the leaflets, which were left with the Antiquities office in both physical and PDF format.

The issue of site information in Swahili was difficult to address during our time in Kilwa, but again this is something that has been pursued later. A recent University of York impact award paid to produce a book '*Kugundua historia ya Kilwa/Discovering Kilwa's History*', written in Swahili and in English and outlining the history and archaeology of the region at a level appropriate for teaching in schools. Copies of the book have been delivered to all schools in the region and are available at the Kilwa Antiquities office. The book can also be read on a phone or tablet and is freely available at www.kilwa.org/.

11.4.4 Community Heritage at Kilwa

Our work at Kilwa could not address all the concerns raised by stakeholders in the region and remains a basis for further work rather than a completed mission. At Kilwa, as discussed, it is one of several projects that have sought to incorporate community voices or to use digital technologies. Our work during the 2018 visit was largely based on listening to local requests and assessing how stakeholders were using existing digital resources such as the laser scans of the sites, or the social media resources. We identified some key challenges around sustainability and access for those resources, which we as an academic community must seek to address in the future.

11.5 Reflections

In this final section, we discuss what we learned from our involvement in co-producing networks for community heritage in Tanzania. Did our engagements with local communities in cultural heritage meet the expected standard of creating a structure for co-production? And to what extent did our training activities build capacity among heritage professionals, local NGOs, community groups and local businesses to sustain and enhance heritage resources in our study sites and elsewhere in Tanzania?

Before addressing these questions, it is worth taking a step back to consider some of the characteristics of the project. One of these was that the project involved a much wider group of people than first envisaged. This became a strength. CONCH encouraged and accommodated volunteers and gave them space to be members of the project. The commitment of community partners to the project was initially very positive and allowed CONCH to evolve during its implementation in response to new opportunities and new variables. This has included new project aims, members, places and even the inclusion of new communities into the fabric of our collaboration. It has also permitted participants to better understand the skills possessed by a wide-ranging membership.

However, other factors outside the control of the project have worked against the sustainability of some of the initiatives. The most acute of these was a change in leadership at Uzikwasa leading to the decision to abandon its earlier focus on promoting cultural heritage conservation and protection in Pangani and involvement in the running of the Pangarithi Heritage Centre. Part of this space is now used as a mid-range restaurant, and while the exhibition materials have been retained, this change in use has meant many local residents (including school-age children) no longer feel comfortable visiting the space. Uzikwasa also no longer has a heritage officer. The previous heritage outreach activities, especially the work with schools, are instead now being undertaken by the former heritage officer on his own or by his newly created NGO (Lenga Mbali) for disabled children (Kleinitz et al., 2024). All of the interpretation signs that Uzikwasa installed are now either missing or illegible, and the momentum that had developed for protecting the town's built heritage under the auspices of the Pangani Conservation Task Force (Lwoga 2017a, 2017b) has diminished. On a more positive note, the results of a recent survey of public attitudes to Pangani's built heritage suggest widespread public support at the local level, and a corresponding desire to see greater central and district government involvement, including gazetting the historic core as a conservation area, in securing its conservation and protection (Chami and Mjema, 2024).

Such concerns are well-founded since Pangani's previously well-preserved waterfront has been particularly affected in recent years, as private owners have replaced late 19th-century buildings with modern houses made of glass and concrete, and the local authorities have installed various kinds of street furniture without any consultation with the Department of Antiquities or inclusion of watching briefs and archaeological impact assessments and mitigation measures. Work on a new bridge providing a much-needed replacement of the current car and foot passenger ferry, financed via loans from China, is nearing completion. This too was undertaken without any prior impact assessment work, and while it runs just north of the main town, construction of the new tarred highway linking Pangani with Tanga has already partially impacted the town's old cemetery where many previous European residents are buried. Kilwa is better placed, as it has dedicated heritage infrastructure around the World Heritage Site. Yet the region is similarly affected by international development, with a huge new port financed by Chinese investors at Kilwa Masoko offering economic possibilities that the heritage sites have so far failed to match.

11.5.1 The Spectrum of Co-Production Networks

The creation of co-production networks was a guiding principle of CONCH. It had the explicit intent to create sustainable networks for skills sharing and the co-production of knowledge between researchers, heritage professionals, community groups and other stakeholders such as tourism operators and tourism-related businesses. CONCH co-production was in practice defined by a set of participants and built heritage spaces in Tanzania. Engaging stakeholders with community heritage entailed a chain of events, from planning to activity phases to intellectual work, with a good deal of improvising along the way. Co-production should, therefore, be seen along a spectrum. Throughout the implementation of the project, we reflected not only on who would produce such knowledge but also on what is co-produced and the roles that different actors might assume in different settings, those actors including members of local communities, academics, heritage professionals, schools and local government bodies. For example, the creation of educational resources at Pangani and at Kilwa was principally achieved through education professionals.

Communications and engagement activities through workshops and professional training field schools introduced at the beginning of the project encouraged wide participation of academic partners and collaborators from local heritage organisations (Uzikwasa, Kilwa Culture Centre), national bodies (Antiquities Division, Tanzania) and some community leaders. Our project activities sought to break down perceived knowledge and practice hierarchies often associated with authorised heritage organisations and discourse. This is evident in the archaeological fieldwork at Pangani where local researchers, including the then heritage office for Uzikwasa, gained experience and subsequently played a role in supervising volunteers in test-pitting and writing watching brief reports and where university students contributed their expertise to provide and disseminate archaeological knowledge and skills to primary school students. It is also evident in the creation of heritage resources at Kilwa, where local community responses and concerns were included in the information provided to tourists. Emphasis on community involvement in knowledge creation helped to dissolve the boundaries between groups.

Regular planning meetings with local communities before and after activities took place helped to provide participants with a sense of legitimacy and group identity and allowed some participants to develop sub-specialisations. Through CONCH, local communities and scholars learned technical skills enabling them to more fully participate in the project. Working together was taken as a skill that needed to be learned and practised by all participants. This included developing an ability to negotiate the technical challenges of working in a foreign country for those who came from abroad.

For local residents, their initial activities probably resembled research training more than co-production, because hierarchies between 'professionals' and 'local communities' are deep-rooted, and many participants were new to each other. As time went on, project members learned soft skills, developing group cohesion and willingness to push the project forward and hence co-production. Enskilment helped all members to reinvent their identities in the project. For example, communities who had participated in earlier focus groups or training activities were regularly involved in later phases of activity. In others, notably some of the offshoots of the project, community members have taken a leading role in dissemination activities including writing blogs.

The CONCH project not only aimed to enable local people to undertake heritage activities and engage with coastal Tanzania's built heritage, but also sought to create a new sense of understanding of the history of their areas and on presenting findings to a range of 'publics'. Online resources helped with this. The network's online presence consisted of a project website, and social media outlets (Facebook, Twitter), fully integrated and sustainable beyond the duration of the network (www.conchproject.org/). The book on Kilwa's history and heritage is fully available and downloadable online, as well as being disseminated via schools. Within East Africa access to the internet is still relatively scarce (31.9%), but of this access, a majority (55%) is from smartphones (datareportal.com/reports/digital-2024-tanzania#). Women are significantly more likely to have access to a smartphone than to a computer. All participants in the network contributed to a series of Field Trip mobile app cards which added East African content to the history, art/museum and architecture categories on Field Trip (www.fieldtripper.com/). These information 'cards' highlighted features of interest in the heritage of the region and were linked from other online platforms, allowing better communication with a broader public within East Africa and beyond. Activities that required more focused project work, teamwork and investment, such as geophysical survey, helped encourage closer ties of integration and may, therefore, provide the groundwork for stronger community relationships.

11.6 Conclusions

The CONCH project began with a diverse group of people with a passion for creating networks for community heritage in Tanzania and eventually with a plan to co-produce research on coastal Tanzania's built heritage landscape. One outcome is that we now feel better placed to understand the nature of co-production, how it was and can be realised and the different forms it has taken and might take in the future, even though other outcomes fell short of our initial expectations. Reflecting on the CONCH project and its afterlife provides us also with lessons that will help better engage local communities in cultural heritage on the eastern African coast in the future.

Acknowledgements

We are grateful to the local communities at Pangani and Kilwa for their continued support, enthusiasm, and friendship over the duration of the CONCH project. Particular thanks go to Uzikwasa, Tanzania Department of Antiquities, the National Museum of Tanzania and the University of Dar es Salaam. We are grateful to funders – the AHRC, UK GCRF, the University of York Impact Fund, the University of Cambridge Arts and Humanities Impact Fund and the University of Cambridge – Research England GCRF-QR Fund, and all collaborating partners at the University of York, University of Bournemouth, University of Cambridge, University of Dar es Salaam, Uppsala University, ArchaeoLink and On-Site Archaeology.

References

Bettinson, G. (2024). *Creators not content: community voices and digital technologies at Kilwa*. MRes dissertation, University of York.

Chami, M.F. and Mjema, E. (2024). Local community engagement and gazettement approach in managing and conserving Pangani historic town in Tanzania, *Built Heritage*, 8, p. 31.

Chirikure, S., Ndoro, W., Bugarin., Flordeliz T., di Lernia, S., Ichumbaki, E. and Lwoga, N. (2021). Usable pasts forum: UNESCO and heritage tourism in Africa, *African Archaeological Review*, 38(3), pp. 513–33.

Chirikure, S. and Pwiti, G. (2008). Community involvement in archaeology and cultural heritage management: An assessment from case studies in southern Africa and elsewhere, *Current Anthropology*, 49(3), pp. 467–85.

Chittick, H.N. (1974). *Kilwa: An Islamic Trading City on the East African Coast, 2 vols*. Nairobi: British Institute in Eastern Africa, Memoir 5.

CyArk (2024). *Kilwa Kisiwani*. www.cyark.org/projects/kilwa-kisiwani/overview

Flores de León, R.M., Babere, N.J. and Swai, O. (2020). Implications of cultural heritage in urban regeneration: The CBD of Dar es Salaam, *Revue Interventions économiques*, 63, 2020, URL: http://journals.openedition.org/interventionseconomiques/9171

Garlake, P.S. (1966). *The Early Islamic Architecture of the East African Coast*. Oxford: Oxford University Press, British Institute in Eastern Africa Memoir 1.

Glassman, J. (1995). *Feasts and Riot: Revelry, Rebellion, and Popular Consciousness on the Swahili Coast, 1856-1888*. Oxford: James Currey.

Gramly, R.M. (1981). Archaeological reconnaissance at Pangani Bay, *Tanzania Notes and Records*, 86–87, pp. 17–28.

Heathcott, J. (2013). Heritage in the dynamic city: The politics and practice of urban conservation on the Swahili coast, *International Journal of Urban and Regional Research*, 37(1), pp. 215–37.

Horton, M., Olsen, J., Fleisher, J. and Wynne-Jones, S. (2022). The chronology of Kilwa Kisiwani, AD 800–1500, *African Archaeological Review*, 39, pp. 143–166.

Ichumbaki, E.B. (2016). A history of conservation of built heritage sites of the Swahili Coast in Tanzania, *African Historical Review*, 48(2), pp. 43–67.

Ichumbaki, E.B., and Mapunda, B.B. (2017). Challenges to the retention of the integrity of World Heritage Sites in Africa: The case of Kilwa Kisiwani and Songo Mnara, Tanzania, *Azania: Archaeological Research in Africa*, 52(4), pp. 518–39.

Kabyemela, M., Mugobi, T.I. and Mjema, E. (2024). Determinants of access to marine cultural heritage assets for poverty alleviation in the coastal communities of Kilwa and Bagamoyo in Tanzania, *Journal of Community Archaeology & Heritage*, 11, pp. 1–16.

Khalfan, K.A. and Ogura, N. (2012). Sustainable architectural conservation according to traditions of Islamic waqf: the World Heritage–listed Stone Town of Zanzibar, *International Journal of Heritage Studies*, 18(6), pp. 588–604.

Kigadye, F.S. (2014). Architectural conservation in rapidly urbanising cities: The case of Dar es Salaam, Tanzania, *Les Cahiers d'Afrique de l'Est, The East African Review*, 49, pp. 29–54.

Kleinitz, C., Mjema, E., Chami, M.F., Kizito, S., Lane, P.J., Ntikachika, E., Ndyanabo, A., Kisoma, M., Hartl-Reiter, C. and Bunk, T. (2024). 'Developing collaborative approaches to the (built) "shared heritage" of German colonialism in Pangani, Tanzania', in: M. Falser, G. Horn and T. Mager (eds.) *Monuments and Sites De-colonial! Methods and Strategies of Dealing with the Architectural Heritage of the German Colonial Era*. Berlin: ICOMOS Hefte des deutschen Nationalkomitees, pp. 68–76.

Lane, P. J. (2010). Developing landscape historical ecologies in Eastern Africa: An outline of current research and potential future directions. *African Studies*, 69(2), pp. 299–322.
Lwoga, N.B. (2017a). Local collaboration network in management of built heritage: A study of Pangani Conservation Task Force (PCTF) in Tanzania, *Journal of Cultural Heritage Management and Sustainable Development*, 7(3), pp. 226–43.
Lwoga, N.B. (2017b). Factors influencing local residents' intentions to conserve the built heritage in Tanzania, *Journal of Heritage Tourism*, 12(4), pp. 394–409.
Mapunda, B.B.B. (2018). *Bagamoyo: From a Slave Port to Tourist Destination*. Dar es Salaam: University of Dar es Salaam Press.
Mjema, E. (2015). Maritime Community Settlement History in Pangani Bay, Tanga Coastal Region,Tanzania. Unpublished PhD dissertation, Goethe University, Frankfurt-am-Main.
Mjema, E. (2018). Catastrophes and deaths along Tanzania's Western Indian Ocean coast during the Early Swahili period, AD 900–1100. *Azania Archaeological Research in Africa*, 53(2), pp. 135–155.
Moon, K. and Blanchard, P. (2009). *Pangani Historic Town: Pangani Heritage Conservation Report*. Unpublished report commissioned by Uzikwasa, Pangani.
OECD (2024). DAC-List-of-ODA-Recipients-for-reporting-2024-25 flows. www.oecd.org/dac/financing-sustainable-development/development-finance-standards/DAC-List-of-ODA-Recipients-for-reporting-2024-25-flows.
On-Site Archaeology (2010). *Pangani, Tanga Region, Tanzania: report on an Archaeological Evaluation*. Unpublished report, OSA Report No. OSA 10EV10, October 2010.
On-Site Archaeology (2013). *Pangani, Tanga Region, Tanzania: Report on an Archaeological Evaluation*. Unpublished report, OSA Report No. OSA 13EV10, October 2013.
Ormell, C.P. (1974–1975). Bloom's Taxonomy and the objectives of education, *Educational Research*, 17, pp. 3–18.
Pradines, S. (2004). *Fortifications et urbanisation en Afrique orientale*. Oxford: British Archaeological Reports, S1216 (Cambridge Monographs in African Archaeology 58).
Rhodes, D.T. (2010). *Historical Archaeologies of Nineteenth-Century Colonial Tanzania: A Comparative Study*. Oxford: BAR International Series 2075 (Cambridge Monographs in African Archaeology 79).
Rhodes, D. T. (2016). *Building Colonialism: Archaeology and Urban Space in East Afric*a. London: Bloomsbury Publishing.
Sadik, N.A., Siraj, H., Harrouni, K. and Sdiqui, A. (2023). Swahili architecture: A heritage of the East Coast of Africa, *African and Mediterranean Journal of Architecture and Urbanism*, 5 (2), pp. 15–27.
St Andrews University (2021). *Rebuilding the past through citizen science*. https://impact.wp.st-andrews.ac.uk/curious-travellers-rebuilding-the-past-through-citizen-science/
Swai, O. (2023). 'Diminishing architectural artifacts along the coastal stretch of Tanzania', in: K. Hmood (ed.) *Conservation of Urban and Architectural Heritage-Past*, Present and Future. IntechOpen.
Tanzania Wildlife Authority (TAWA) (2023). *Ruins of Kilwa Kisiwani and Ruins of Songo Mnara World Heritage Site Integrated Management Plan 2023 – 2032*. Dar es Salaam: TAWA.
United Republic of Tanzania (URT) (2008). *Cultural Heritage Policy*. Dar es Salaam: Ministry of Natural Resources and Tourism. Antiquities Department.
Walz, J.R. (2010). Route to a Regional Past: An Archaeology of the Lower Pangani (Ruvu) Basin, Tanzania, 500-1900 CE. Doctoral dissertation, University of Florida.
Wamboye, E.F., Nyaronga, P.J. and Sergi, B.S. (2020). What are the determinants of international tourism in Tanzania?, *World Development Perspectives*, 17, pp. 100–175.

Wynne-Jones, S. (2018). 'Kilwa Kisiwani and Songo Mnara', in: S. Wynne-Jones and A. LaViolette (eds.) *The Swahili World*. London, Routledge, pp. 253–259.

Wynne-Jones, S., & Fleisher, J. (2015). Conservation, community archaeology, and archaeological mediation at Songo Mnara, Tanzania. *Journal of Field Archaeology*, 40(1), pp. 110–119.

Wynne-Jones, S. and A. LaViolette. (2018). *The Swahili World*. London: Routledge.

Zamani Project (n.d.). Monuments of Kilwa Kisiwani, www.zamaniproject.org/site-tanzania-kilwa-kisiwani.html

Part III

Latin America

12 Santa Cruz Mixtepec

Community Engagement and Religious Heritage in Mexico

Alanna Cant

12.1 Introduction

This chapter discusses community engagement in heritage development at an active religious site, Santa Cruz Mixtepec, in the state of Oaxaca, Mexico, placing it within the broader historical and economic contexts of heritage and tourism in the region. Santa Cruz is the site of a ruined 16th-century Dominican convent (*exconvento*), of which the chapel is still in use as the main church for the local Catholic parish, also called Santa Cruz Mixtepec. From 2015 to 2018, the cloister roof, walls and sections of the refectory were conserved and partially restored by heritage professionals through a collaboration between the municipal government of Santa Cruz Mixtepec, the Mexican federal government via the National Institute of Anthropology and History (INAH) and the Alfredo Harp Helú Foundation of Oaxaca (FAHHO), a privately funded cultural and public health organisation that is maintained by one of the wealthiest families in Mexico.

Notably, the Catholic archdiocese and the parish priest were not included as formal project partners. This was the result of historical circumstances relating to the designation of churches as national property in Mexico and to the legal arrangements that grant Indigenous communities like Santa Cruz Mixtepec a high level of autonomy in Oaxaca. While on one level, this autonomy enables communities to choose whether or not to participate in heritage projects, the legislative frameworks that define churches as property of the officially secular state can also result in mismatched priorities and outcomes.

Before discussing the conservation project and the benefits and challenges it presents in terms of community engagement, I introduce the site and its historical and cultural setting, placing the project within the larger context of Mexican heritage and Oaxacan cultural tourism today.

12.2 Historical and Social Context

Santa Cruz Mixtepec is a relatively small municipality located in the Zimatlán District of Oaxaca, about 55 kilometres south of the state capital Oaxaca de Juarez (not to be confused with the Mixtepec area of the Sierra de Miahuatlán in

DOI: 10.4324/9781003491071-15

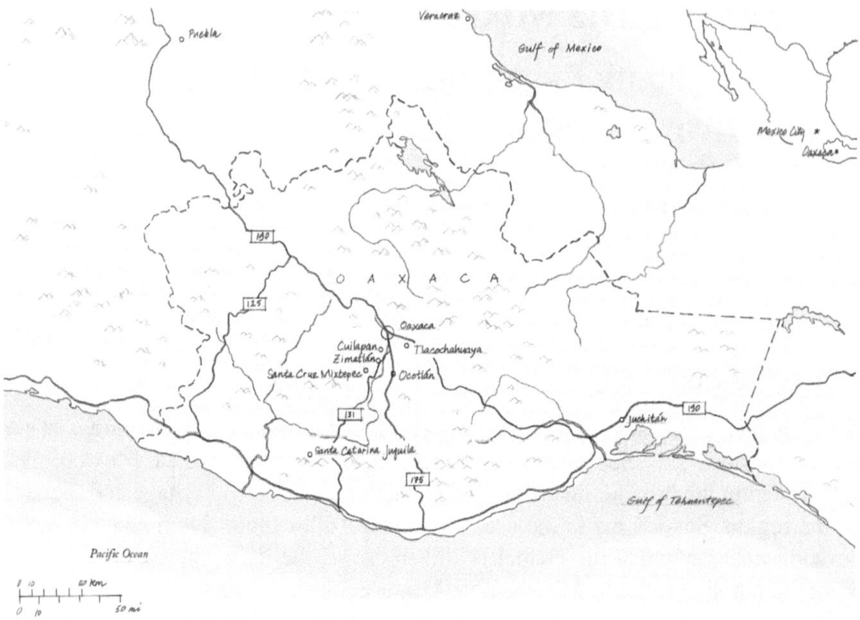

Figure 12.1 Map of the state of Oaxaca, showing the location of Santa Cruz. Mixtepec. © Molly Halloran.

NW Oaxaca; Figure 12.1). It is located in the Central Valleys region of the state, which is historically and culturally associated with the Bën za (Valley Zapotec) Indigenous people, whose language forms one of the four basic Zapotec language groups (the state of Oaxaca has 16 distinct Indigenous languages, with significant linguistic variation in each). The current name 'Mixtepec' is Nahuatl rather than Zapotec, dating to the mid-15th century when the region was conquered by the Aztec Empire, during the reign of Moctezuma I (1440–1469) (Isaac 1983); Mixtepec appears to be the translation of the original Zapotec placename, 'Danicahue': both are toponyms meaning "hill among the clouds" (Pérez Cruz et al., 1991, p. 83).

Nestled in the foothills of the Sierra Madre del Sur mountain range, Santa Cruz and its three neighbouring municipalities form a cultural and ecological transition zone between two distinct regions: the arid fields and rolling hills of the Central Valleys and the temperate humid forests of the Southern Sierra mountains. The village proper of Santa Cruz Mixtepec sits within small valleys and steep ravines at the confluence of the San Juan and San Bernardo rivers, the latter of which drains into Río Atoyac, the largest river in the region, about 5 kilometres south of the municipality. The village is surrounded by high hills that have numerous pre-Hispanic Zapotec occupation and burial sites. Although there have been no excavations, a surface survey conducted in the 1970s of Cerro de la Cueva (Hill of the Cave) on the northwest edge of the village indicates an extensive palace complex. It includes several classic Mesoamerican architectural features, such as platforms, patios, a ballcourt and a significant number of tombs, in which pots and other vessels were found. The earliest occupation at this site is between 100 BCE and 100 CE [Monte Alban II], with the principal occupation from 200 to 500 CE [Monte Alban IIIA] (Winter 1978). It is likely that the other sites in the Mixtepec area share this chronology. These archaeological sites have not been designated as formal 'archaeological zones' by the Mexican federal government, but they fall under the protection of the Federal Law for Archaeological, Artistic and Historical Zones and Monuments, 1972.

Because of its established Zapotec population, Mixtepec became one of the first sites of Christian evangelisation in this part of what became New Spain. The Dominicans initially had exclusive permission to evangelise the Zapotec peoples, granted in 1536 by the infamous conquistador Hernán Cortés, who was by then the Marquess of Oaxaca (Martínez Sola, 1998). Dates attributed to the convent's establishment vary. The records of the Dominicans in Oaxaca first mention Santa Cruz in 1556, when it was recorded that there were ten friars living there (Mullen 1975, p. 235). By 1557, Santa Cruz was incorporated into the Diocese of Antequera and the friars had already established a mill and were developing the local economy based on agriculture, timber, and mining (Burgoa, 1934 [1674], p. 24–25; Mullen, 1975, p. 234–237; Pérez Cruz et al., 1991, p. 83–84). However, by the 17th century, they were losing influence in the region due to epidemics of European diseases, which killed up to 1.35 million people in Oaxaca in less than 100 years, and because the Spanish Crown increased its control over the mendicant orders due to worries about their allegiance to the Pope (Norget, 2008, p. 145; Romero Frizzi, 2000, p. 310–312).

By the end of the War of Independence (1810–1821), the convent in Santa Cruz was abandoned by the friars, enabling wealthy outsiders to establish an hacienda and sugar mill (*trapiche*), the ruins of which can still be seen today. In 1859, the last Dominicans left Oaxaca, after the new republic's government suppressed and expelled the male Catholic orders across the country (Morales, 1998; Torres Torres, 1997, p. 338). By the end of the Mexican Revolution (1910–1920), the convent site was in ruins. It was further damaged by a large earthquake in January 1931, and in the summer of 1936, it was occupied by the Mexican federal army in its attempts to put down an insurrection by Catholic partisans against the new anti-clerical government; in November of that year, the community was taken by Juventino Montaño, the self-proclaimed 'brigadier general' of the Oaxacan Catholic rebels (García Morales, n.d.; Meyer, 2007, p. 194–195). As such, Santa Cruz Mixtepec directly speaks to the three most significant periods of Mexican history, in terms of contemporary perceptions of the country's national identity and heritage: the Mesoamerican cultures of the pre-Hispanic period, the colonial period of New Spain and the Mexican Revolution and its aftermath (Benjamin, 2000; Bueno, 2016; Salas Landa, 2024, p. 107–137).

Today, about 1,200 Cruzeños (people from Santa Cruz) live in the village proper of Santa Cruz Mixtepec, which serves as the administrative centre (*cabecera municipal*) of the larger municipality of about 3,700 people (INEGI 2020); 22% of the population speaks Zapotec in addition to Spanish, which is the primary language used outside of the home in this region (INEGI 2020). Santa Cruz is also the centre of the Roman Catholic parish, which includes the neighbouring but independent municipalities of San Miguel Mixtepec, San Bernardo Mixtepec and Magdalena Mixtepec. Although each village and hamlet in the parish has its own church or chapel, most of which date to the colonial period, they are all served by the single parish priest, who is based in Santa Cruz.

The vast majority of Cruzeños identify as Roman Catholic, although two small Evangelical Protestant churches are also present and active (INEGI, 2020). As in other communities across Oaxaca, Roman Catholic Cruzeños practise their religion through a dynamic combination of 'official' and so-called popular practices and beliefs (Norget, 2006; 2021). They most regularly engage with the Church through Mass, catechism and sacramental rituals, directed by the parish priest; some travel to the larger church in Zimatlán for other prayer and social activities, such as rosary groups. Weekly Mass is well attended by men, women and children of all ages, and the sacraments of baptism, confirmation and marriage are especially important to their social identities and standing in the community. The celebration of many saints' days, especially the large *fiesta* on September 14th for the Exaltation of the Holy Cross (the village's patron), includes a special Mass and procession led by the priest. At the same time, Cruzeño Catholics engage with the sacred and supernatural through activities and beliefs that stand outside of the official teachings of the Church. These include devotional, penitential and magical activities directed to saints' images both inside churches and on home and workplace altars, pilgrimage to sacred places in the local landscape and further afield and the observance of religious traditions not officially recognised by the Church, but which build on normative Roman Catholic beliefs and practices (Cant, 2018a; cf. Norget, 2021). A noteworthy example is the making of household altars and collective celebrations/commemorations in and around the municipal cemetery for the Day of the Dead (*Día de Muertos),* which takes place over the annual Catholic feasts of All Saints and All Souls, on November 1 and 2 (see Norget, 2006 for a comprehensive account and analysis).

The parish church itself, which was the main chapel of the colonial convent, is at the heart of all Catholic activities in Santa Cruz, even those that fall outside of the official Church's remit (Figure 12.2). Access, physical maintenance and the organisation of services and other activities are coordinated by the parish church committee, which is overseen by the municipal authority.

Although it may seem odd that the church is controlled and managed by parishioners rather than the priest, this is in part the result of traditional Indigenous governance systems in Oaxaca (*usos y costumbres*), which are centred upon communal decision-making through citizens' assemblies and public service roles (*tequio*) undertaken by married men and, in some cases, women (Canedo Vásquez, 2008). In Santa Cruz, positions on the church committee are included in the formal governance system, which had practical and political implications for the convent

Figure 12.2 Exterior of the front (west side) of Santa Cruz Mixtepec. © Alanna Cant.

heritage project (see below). Since 1995, these systems have been protected by the Oaxacan state constitution, which guarantees that in the municipalities where such systems are used (418 of 570 total), communities' self-determination is legally protected from outside influence by the state and political parties, and in effect from other organisations and corporations, sometimes including the Catholic Church itself (Anaya Muñoz, 2005; Magaloni et al., 2019). Additionally, ownership of the parish church lies with the Mexican federal government, not the Catholic archdiocese, a point I return to below.

12.3 The *Exconvento* of Santa Cruz Mixtepec

The religious complex of Santa Cruz Mixtepec is set within a large walled atrium or yard, which forms one side of the village's central plaza. The main structure is made of locally quarried stone and lime mortar. It has a small chapel (43.5 m × 9.2 m) of a single nave with no transepts (Figure 12.3). The nave is oriented east to west, with the chancel/presbytery and main altar on the eastern end, as was typical of this period. The chancel includes a small rectangular apse, which is separated from the nave by a triumphal arch. The apse contains a 17th-century neoclassical wooden reredos (*retablo*) altarpiece, which obscures remnants of 16th-century murals on the wall behind. The western end likely had a choir, as indicated by

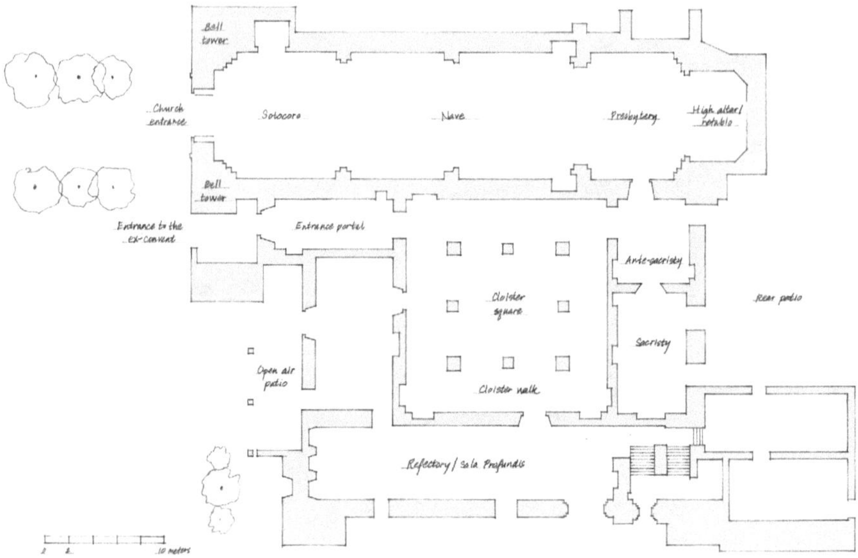

Figure 12.3 Floorplan of the existing structures at the Convent of the Holy Cross. © Molly Halloran.

the vestiges of arches that can still be seen high on the interior wall. The ceiling has barrel and domical vaults, the roof has two tiled domes, and there are two belfries on the western end. Also on the western end is the arched entrance to the church, which includes an 'alfiz' (rectangular moulding around the arch) and Dominican fleur-de-lis motifs (see Figure 12.2; the domes and belfries were likely added in later stages of construction) (Mullen, 1970, p. 68–74; 198; FAHHO, 2022, p. 45–46).

The southern wall of the nave is shared with the convent area of the site, and there is a door between them in the chancel end of the nave. Embedded in this very thick shared wall is a confessional, where penitents would have entered from the church and the confessor from the convent side (see Mullen, 1970, p. 68). The convent also has an entrance on the western external façade, to the right of the chapel's entrance (Figure 12.2). The central part of the convent was two stories high and centred around a small, arcaded cloister, which had beamed wooden ceilings, red adobe brick floors and plastered walls with niches and very fine murals. Other spaces in the convent included the ante-sacristy and sacristy, refectory, kitchen, dormitories and an outdoor portico or porch for pilgrims, which today are in various states of disrepair (FAHHO, 2022, p. 46; Figure 12.3). To the south and east of the main site, there are also the undated ruins of further outbuildings associated with the convent, which were sold off to local people before and after the property was taken over by the state (see below).

12.4 Heritage in Mexico and Cultural Tourism in Oaxaca

To understand the development of the Santa Cruz Mixtepec project and the relationship between the community and the other parties involved, it is necessary to provide a brief sketch of the broader context of heritage in Mexico and cultural tourism in Oaxaca.

Mexico has a long history of heritage protections, which effectively began with a series of regulations governing pre-Hispanic sites and the collection and export of artefacts in the 19th century (Bueno, 2010). The first comprehensive heritage law was passed in 1939 with the incorporation of INAH by the post-Revolutionary government; an average of nine national laws and amendments pertaining to heritage have been passed in each decade since then (UNESCO, 2023).

As has been extensively documented by scholars, heritage became one of the key pillars used by the government and intellectuals in the mid-20th century to repair the social fabric and foment a new national identity after the horrific violence and divisions caused by the Mexican Revolution (Coffey, 2012; García Canclini, 1993; Knight, 2015; Rozental, 2016). Pre-Hispanic archaeology and contemporary Indigenous arts and cultural forms played an important role in this throughout the 20th century, as they were seen as the 'deep roots' of indigeneity upon which the new mestizo national identity was built (Alonso, 2004; López, 2010). At the same time, colonial architecture such as town squares, theatres, villas and churches became central to this political project, often with only their exterior aesthetics preserved for their heritage and touristic value, as many were repurposed for government offices, schools, museums, hospitals, et cetera (Lira Vásquez, 2014, p. 73).

Churches and former convent buildings were especially suitable for such 'heritagisation', firstly because many had suffered significant material damage during the Revolution and its aftermath: their large sizes and strategic positions had made them ideal barracks and headquarters for different factions (Knight, 2015, p. 317). Through their restoration and repair, the new national government could reframe these overt symbols of colonial rule into symbols of legitimacy for the Republic (Knight, 2015, p. 317). Secondly – and perhaps more importantly for the case here – the constitution of 1917 rendered all churches property of the federal government, rather than the Catholic Church (Bantjes, 2007, p. 112–113). This was initially part of the broader agrarian reforms and anticlerical politics of both the early independent and post-Revolutionary republican governments, but it remains enshrined in law to this day: all church properties (of all denominations) built before 1992 are federally owned as 'national patrimony', and therefore, the state, and not the Church, has control and is legally responsible for them.

As early as the 1930s, the celebration of pre-Hispanic sites, colonial architecture and contemporary Indigenous cultures in the emerging national imaginary enabled Oaxaca to develop as a culture and heritage destination for tourists from Mexico City and the United States (Lira Vásquez, 2014, p. 73). Boosted by strong government sponsorship and marketing for its traditional and Indigenous crafts, foods, cultural activities and archaeology, Oaxaca became dependent on its 'economies of culture' by the end of the 20th century, increasingly marketing itself as the place to experience the 'most authentic Mexico' (Cant, 2019, p. 8–14; Poole, 2004). While the state has several small but popular beach destinations, investment and marketing continue to be focused on the promotion of tangible and intangible heritage in the state capital and throughout the Central Valleys region. As anthropologist William Wood has observed, Oaxaca's photogenic character is not the spontaneous effect of ordinary people going about their daily lives: the Oaxacan state has gone to great lengths to produce, control and disseminate an image of picturesque authenticity through marketing, urban planning, and by directing investment towards specific types of businesses and organisations (2008, p. 32; Lira Vásquez, 2009). Indeed, the belief in Oaxaca's cultural uniqueness continues to spur tourism today: the past decade has seen exponential growth in this market, with the development of traditional and very high-end culinary tourism, focusing especially on mezcal, the currently very trendy agave spirit of Oaxaca's countryside (Brulotte, 2017; Brulotte and Starkman, 2014).

Although Oaxaca's cultural and heritage tourism may appear to be less commercial than the large beach resorts elsewhere in the country, it is in fact very big business. In 2018, the state registered over 5 million tourist arrivals, generating around 15 billion pesos (ca. £705 million) of revenue (SECTUR, 2018), and there are indications that tourism in 2024 will exceed pre-COVID-19 levels (Valadez, 2024). Mexicans make up the largest percentage of tourists, a majority of whom are from Mexico City and the larger regional towns of central and northern Mexico. They are especially drawn to Oaxaca during the two most important religious/cultural holidays, Holy Week and Day of the Dead. Americans, Europeans and others have always made up a significant number of arrivals, too, and have typically spent longer (and more money) in the region, sightseeing, visiting craft-making villages, regional marketplaces, et cetera (see Cant, 2019; Wood, 2008). However, the shift towards culinary and spirits tourism over the past decade has now attracted younger, wealthier Americans and Europeans to the state, with many increasingly searching for business and investment opportunities in mezcal, in addition to an exciting holiday or opportunities for social media careers.

Despite the growth and the very visible influx of tourists and cash into the capital and those communities that profit from the new mezcal market, the benefits of the tourist economy remain in the hands of the few: Oaxaca is characterised by high levels of poverty and inequality. The 2020 national census shows that 66% of Oaxacans live in either 'moderate' or 'extreme' poverty, and only 8.5% are classed as living without poverty or another vulnerability (CONEVAL, 2020, p. 13–14). In Santa Cruz Mixtepec itself, 81% of people live in 'moderate' or 'extreme' poverty (INEGI, 2020). In such a context, it is not at all surprising that Santa Cruz was interested in improving the condition and profile of the convent ruins: not only would it support the preservation of a site that is extremely important to their own sense of community and identity (Cant, 2024a), but it could potentially allow the village to attract day tourists from the capital or those passing along the highway to the coast, seemingly offering economic opportunities for small businesses – especially since traditional mezcal production also takes place in Santa Cruz and the neighbouring village of San Bernardo Mixtepec. Tourism is a key area of development in the most recently published Municipal Development Plan (Santa Cruz Mixtepec, 2020, p. 114–115).

12.5 The Santa Cruz Mixtepec Project and Its Formal Stakeholders

Initial surveys conducted by engineers and conservationists from the FAHHO's Restoration Workshop identified several areas of concern with regards to the condition of the material fabric and maintenance of the ruined convent. The most significant problems were in the cloister: ever since the second storey disappeared, quite possibly in the 1931 earthquake, the ground level has been exposed to the elements. This was causing structural deterioration, staining and vegetation growth on the walls and adobe floor, as well as greatly damaging the remnants of the 16th-century murals on the lime plaster walls, which showed "partial detachments [of pigments], stains caused by humidity and loss and degradation of pictorial strata" (FAHHO, 2022, p. 50, translation by author; Figure 12.4). Twentieth-century additions to the building, including reinforcements for concrete walls and a metal spiral staircase, which had been drilled into the original 16th-century walls, were also causing structural issues (see Figure 12.5). Other rooms, such as the refectory and those adjacent to it, were in a bad state of repair – in 2017, they were impacted by the

Figure 12.4 The murals in the cloister, after conservation. © Alanna Cant.

Figure 12.5 Interior of cloister during conservation. The staircase can be seen on the right. © Alanna Cant.

very large earthquakes that struck central and southern Mexico, and they likely would have fallen in the near future had the project not already been underway (see Cant, 2018b, for more about these earthquakes and their effects on religious heritage).

To address these issues, a formal collaborative partnership was undertaken between three stakeholders: (1) the Restoration Workshop of the FAHHO; (2) the community of Santa Cruz Mixtepec, through its municipal authority and the church committee; and (3) the Mexican Federal Government, via INAH.

The project was conducted in four distinct phases: informally from 2013 to 2014 and then under three separate formal agreements from 2015 to 2018 (FAHHO, 2022, p. 44). In 2013, members of Santa Cruz's church committee were directed by the municipal president to attend meetings and grant development workshops run by the FAHHO Restoration Workshop, and they began to undertake research and fundraising to support their application. The project was officially announced in 2015, after it won the required federal permission and grant, with work beginning on site in 2016. The funding for the project was subsequently renewed for the second and third stages in 2017 and 2018. Although there was still a significant amount of work on the site that could have been done after the third stage, projects of this size rarely receive funding for more than three rounds. The total cost was approximately 2.2 million pesos (ca. £104,000), with 1 million pesos (ca. £39,000) from the FAHHO, 800,000 pesos (ca. £31,000) from the federal government via a grant from the Secretariat for Culture (Secretaría de Cultura) and 410,000 pesos (ca. £16,000) from fundraising by the citizens and municipality of Santa Cruz Mixtepec (FAHHO, 2022, p. 44). No financial contributions from the Archdiocese of Oaxaca were permitted.

Founded in 2005, the FAHHO's Restoration Workshop provides technical support and expertise, project and grant management, and funding for the conservation of Oaxaca's built heritage (see FAHHO, 2024). As part of the larger 'family' of the FAHHO, which includes museums, cultural institutes and programmes, research libraries and archives, children's libraries and centres and a baseball academy, it seeks to enact its founder Alfredo Harp Helú's declared principle of apolitical philanthropy (see Gazzo Reisman, 2022). However, some observers have expressed concerns about the personal power of the Harp Helú family and their control over such a high number of cultural institutions and historic buildings in Oaxaca. Drawing on the wealth and connections of its founder, an extremely wealthy businessman from one of the most well-known elite families in Mexico, the FAHHO has been accused by some of shepherding in gentrification and shoring up plutocratic power in Oaxaca (see Rodríguez, 2014)

As mentioned above, the community partners in the convent project were primarily members of the church committee, who did this work as *tequio,* or unpaid public service, within the traditional governance system of their community. The church committee is accountable to the municipal authority and president, who in turn are accountable to the entire community via the citizens' assembly process. As work on national heritage properties requires permits, oversight, reporting and the engagement of professionally trained conservationists and architects, the project required a large amount of start-up time and work from the church committee in addition to their regular responsibilities. While capacity-building activities are provided by the Restoration Workshop, most members of the church committee did not have post-secondary education or relevant experience that would help them to navigate the complexities of the paperwork and planning, although one member was trained as an electrical engineer, which helped the committee enormously.

Apart from funding and permits, the federal government was more removed as a partner in the project itself, although INAH's Oaxaca office provided expertise from art historians and the conservator who worked on site for three months to stabilise the murals in the cloister. As INAH and the Restoration Workshop work together on over twenty projects per year in Oaxaca, their staff know one another very well, and they work together to coordinate with communities and support the projects that are funded through these partnerships.

12.6 Community Engagement: Benefits and Challenges at a Religious Heritage Site

As is clear from the discussion above, the Santa Cruz Mixtepec project included a huge amount of community engagement from the outset. This enabled the project to successfully deliver the three significant stages of work that have greatly improved the condition of the convent site and ensured its future for the community and the wider public. Although the realities of the traditional governance systems mean that such projects would be effectively impossible without community agreement, all the heritage experts were genuinely interested in working with community partners and co-producing project proposals, budgets and plans. The engineers, conservators, archaeologists and architectural historians of the Restoration Workshop have become experts in communication and engagement, as they assist communities in navigating the complex landscapes of research, permits and grant applications that enable them to undertake their work of restoration and conservation across the state. This is the result of almost twenty years of institutional development and learning on their part.

An important reason the FAHHO requires communities to contribute a portion of the project costs is that it increases 'buy-in' and a sense of responsibility towards the project by those beyond the church committees and municipal authorities; this is particularly so as most need to fundraise through community networks to meet their contribution targets (this also helps increase 'buy-in' from former residents who have emigrated, who often have more funds available to contribute). The importance of the commitment from the entire community is illustrated by a project in another region of the state that came to a halt when the municipal president's term came to an end; the new president rallied the citizen's assembly to withdraw its support, leaving the project uncompleted and the funds wasted.

There are also cases in which the heritage values of historic churches become exploited in larger political issues between autonomous communities who have conflicts over land or between communities and the federal government: the legal status of churches as 'national patrimony' can make them a focus for expressing frustration and disagreement. In one case a community expelled some heritage inspectors from the federal government, closed its borders and tore down its historic church: building a new church from scratch was far cheaper than the costs associated with restoring a 16th-century building. As many hundreds of communities across Oaxaca have historic churches, there is no shortage of opportunities for the Restoration Workshop to develop projects, so there is no need to work on projects where they are not wanted. But it is essential that once a project has begun, they do everything possible to ensure that community support is maintained, especially with autonomous municipalities where leaders have the capacity to expel non-residents from their territories.

Despite these efforts and the quite effective systems of community engagement that are now in place, when it comes to religious sites the strongly secular (and even anti-Church) foundations of the Mexican state, heritage legislation and practices of heritage experts can result in more subtle forms of disconnect between heritage values and communities' own values and desired outcomes. Of course, those heritage professionals who have spent sufficient time living and working in rural communities have knowledge and respect for the so-called popular Catholic practices described above. Additionally, recent discussions about professional ethics in the heritage sector, including higher education, museums, and academic research in Mexico have encouraged better community engagement and pluralism in the treatment of objects and spaces, especially where these are central to local identities and local spiritual practices. However, as much as these experts approach community members as partners, their primary professional and legal responsibilities remain firmly rooted in secular heritage, conservation and art historical values (INAH, 2016). Although we must not assume that individual heritage experts themselves are not religious (78% of the population in Mexico identifies as Roman Catholic [INEGI, 2020]), the nature of their work does not allow them to fully take into account particular spiritual perspectives when it comes to decision-making, valuation, interpretation or planning at religious heritage sites (although this is a little more fluid in projects that include 'intangible heritage' components).

This tension was most evident in Santa Cruz Mixtepec when it came to the identification and treatment of historic statues of saints, which are understood by many Catholics as actually embodying the saint that they represent, through the Catholic doctrine of immanence. One statue was thought by the villagers to be San Jacinto (St Hyacinth of Poland), which they believed was an object of significant historic and religious value. San Jacinto had been a holy patron of the community when it was first founded, and the statue itself was considered to be spiritually active and particularly important to local perspectives on identity, history and placemaking (see Cant, 2024a). The heritage experts, on the other hand, dismissed it as just another colonial statue of San José (St Joseph), which are exceedingly common in Mexico. The upshot of this was that there was no official support for the conservation of the statue, which was not in good condition, and it was not included in any of the project stages, even though 'moveable heritage' could have been included in the project plan. The community responded by fundraising to contract a 'black market' (i.e. self-taught and uncertified) art conservator to undertake the work, as they felt the repairs were essential for themselves and their relationship with San Jacinto and other saints.

A second issue is the way that 'community' is defined in relation to religious heritage and the assumptions that are built into processes about who is considered a rightful partner and stakeholder in the first place. The Restoration Workshop always approaches community engagement by building relationships with municipal authorities. This makes logistical and political sense, as they really do need the cooperation of the authority for the project to even get off the ground. However, in Santa Cruz Mixtepec, there are other groups that could rightfully be considered part of the 'community' of the religious site and have very real interests in the outcomes and funding of the project but were not directly involved or consulted. The most obvious of these is the parish priest and the Roman Catholic Archdiocese of Oaxaca. Although the expropriation of Mexican churches by the state in the 19th and 20th centuries means that the Church no longer owns them *as property*, they were not deconsecrated and they remain used as places of worship and are considered sacred spaces. The deep sacredness of church buildings for Catholics is due to the presence of Jesus Christ in the transubstantiated bread and wine of Mass and to the spiritual consecration of the *entirety* of church buildings, not just the chancel, altar, bible or holy objects. For Catholics, the entire material fabric of their religious heritage is fundamentally sacred and intrinsic to their spiritual beliefs and practices (Cant, 2024b; cf. Gilchrist, 2020).

A second group that was not consulted about the project were Catholics who do not live in the municipality of Santa Cruz Mixtepec, but in the neighbouring villages covered by the Roman Catholic parish (see above). The church in Santa Cruz is their main parish church and is where many attend weekly mass and visit their priest for confession or other needs. Although it is likely that they would be happy with the project overall, they do not have representation on Santa Cruz's church committee, as they complete their public service *tequio* in their own municipalities and participate in their own, separate, citizens' assemblies.

Finally, a third group that the project did not engage with is those residents of Santa Cruz Mixtepec municipality who are not Catholic. Because only Catholic Cruzeños would accept *tequio* on the church committee, and Catholics are a very large majority in the village, the municipal authority and church committee only effectively represent Catholic villagers' perspectives and opinions. One man who belonged to one of the small Protestant churches in the village told me that he was unhappy with the amount of time, energy and resources that the municipal authority directed towards Catholic fiestas and activities and that the conservation project was a particularly conspicuous example of this. Although he recognised that the site was historically significant to the community and its identity, he and his family would not even enter the grounds of the church and the convent for religious reasons. Although in principle he could have raised these concerns at the citizens' assembly before the project was initiated, or during the subsequent stages of renewal, non-Catholics often experience discrimination in these communities organised around traditional governance, especially when they decline to participate in particular *tequio* roles. Punishments can include the withholding of public services, such as water and electricity, and in extreme cases, families might be jailed or expelled from the community altogether (e.g. Mundaca, 2023; Rodriguez, 2021).

12.7 Reflections

The case of the religious heritage at Santa Cruz Mixtepec shows the complex interplay of legal, social, economic and religious dynamics, which should be taken into account when considering community engagement with religious heritage projects. Although Mexico is a particularly good example of how entrenched ideas about religion and community can become woven into current frameworks of heritage development and tourism, the case can serve to highlight particular features that should be considered in religious heritage projects more generally.

The collaboration between the FAHHO and Santa Cruz's municipal authority highlights both the potential and the challenges of engaging communities in religious heritage. The clear benefits were that the project developed in ways that furthered the desires of the community to improve the condition of the convent site, hopefully making it more appealing to visitors as Oaxaca's tourism economy continues to grow. This also benefited the FAHHO and the broader heritage sector, as the material fabric of the site is now secure and its future is ensured.

However, the project was not without points of disconnection, which indicate areas of friction within the overall concept of community engagement with religious heritage in this context. In particular, the prioritisation of secular heritage values over the spiritual needs of the Catholic community and the Catholic Church highlights a broader issue about what and whom churches are for in the 21st century. At the same time, assumptions that everyone in a local community is necessarily of the same mind, or that communal structures effectively represent everyone who lives there, can lead to some people being excluded from projects or not having their objections sufficiently considered. Nevertheless, the example of the FAHHO's Restoration Workshop shows how heritage organisations can be pro-active and flexible when engaging with communities by supporting them through capacity-building and with their heritage and project management expertise.

Acknowledgements

The research for this chapter was conducted in 2017. It was funded by the Marie Skłodowska-Curie Actions (MEXRES 701601) of the European Commission. The research ethics and informed consent procedures were approved by the Marie Skłodowska-Curie Actions selection committee (2015), and the Research Ethics Committee of the School of Anthropology and Conservation, University of Kent (2017).

References

Alonso, A.M. (2004). Conforming Disconformity: 'Mestizaje,' Hybridity, and the Aesthetics of Mexican Nationalism, *Cultural Anthropology,* 19(4), pp. 459–490.

Anaya Muñoz, A. (2005). 'The Emergence and Development of the Politics of Recognition of Cultural Diversity and Indigenous Peoples' Rights', in: *Mexico: Chiapas and Oaxaca in Comparative Perspective*, *Journal of Latin American Studies* 37(3), pp. 585–610.

Bantjes, A.A. (2007). 'The Regional Dynamics of Anticlericalism and Defanaticization in Revolutionary Mexico', in: M. Butler (ed.) *Faith and Impiety in Revolutionary Mexico.* Basingstoke: Palgrave Macmillan, pp. 111–130.

Benjamin, T. (2000*). La Revolución: Mexico's Great Revolution as Memory, Myth and History*. Austin: University of Texas Press.

Brulotte, R. (2017). Alcohol and Ambivalence in Mexico and the Americas, *Latin American Research Review,* 52(5), pp. 910–915.

Brulotte, R. and Starkman, A. (2014). 'Caldo De Piedra and Claiming Pre-Hispanic Cuisine as Cultural Heritage', in: R. Brulotte and M.A. Di Giovine (eds.) *Edible Identities: Food as Cultural Heritage*. London: Taylor & Francis, pp. 109–124.

Bueno, C. (2010). Forjando Patrimonio: The Making of Archaeological Patrimony in Porfirian Mexico, *Hispanic American Historical Review,* 90 (2), pp. 215–245.

Bueno, C. (2016). *The Pursuit of Ruins: Archaeology, History, and the Making of Modern Mexico*. Albuquerque: University of New Mexico Press.

Burgoa, Fr. Francisco de (1934) [1674]. *Geográfica Descripción* Part 2, Vol. 2. Ciudad de México: Archivo General de la Nación.

Canedo Vásquez, Gabriela. (2008). 'Una conquista indígena. Reconocimiento de municipios por usos y costumbres en Oaxaca (México)', in: Alberto Cimadamore (Comp.) *La economía política de la pobreza*. Buenos Aires: CLACSO, pp. 401–426.

Cant, A. (2018a). What Can Witchcraft Do in Mexico? *Anthropology News*, October 28, 2018.

Cant, A. (2018b). Saints of the Shaking Earth. *The Religion Factor*. Available at: www.rug.nl/research/centre-for-religious-studies/research-centres/centre-religion-conflict-globalization/blog/saints-of-the-shaking-earth-10-01-2018.

Cant, A. (2019). *The Value of Aesthetics: Oaxacan Woodcarvers in Global Economies of Culture*. Austin: University of Texas Press.

Cant, A. (2024a). Material Historicity: Rethinking Objects Through the Mis/Recognition of a Mexican Saint, *History and Anthropology*, pp. 1–25.

Cant, A. (2024b). What a New Plan to Save the UK's Churches Says About Their Purpose in Society, *The Conversation,* February 7, 2024. Available at: https://theconversation.com/what-a-new-plan-to-save-the-uks-churches-says-about-their-purpose-in-society-222782.

Coffey, M. (2012). *How a Revolutionary Art Became Official Culture Murals, Museums, and the Mexican State.* Durham: Duke University Press.

CONEVAL (Consejo Nacional de Evaluación de la Política de Desarrollo Social) (2020). *Informe de Pobreza y Evaluación 2020 – Oaxaca*. Available at www.coneval.org.mx/coordinacion/entidades/Documents/Informes_de_pobreza_y_evaluacion_2020_Documentos/Informe_Oaxaca_2020.pdf

FAHHO (Fundación Alfredo Harp Helú Oaxaca) (2022). *Memorias de Restauración* 10(16) Enero – Junio 2022. Available at: https://tallerderestauracionfahho.org/wp-content/uploads/2023/01/CONVENTOS-WEB.pdf

FAHHO (Fundación Alfredo Harp Helú Oaxaca) (2024). *Taller de Restauración FAHHO* website. Available at: https://tallerderestauracionfahho.org/.

García Canclini (1993). 'Los Usos Sociales Del Patrimonio Cultural', in: E. Florescano (ed.) *El Patrimonio Cultural De Mexico*. Mexico City, Mexico: Fondo de Cultura Economica, pp. 41–61.

García Morales, P. nd. Unpublished manuscript (2002). *Escrito de sus recuerdos de Santa Cruz Mixtepec.*

Gazzo Reisman, L.M. (2022). Local Philanthropy and the Transformation of Culture in Oaxaca, Mexico, *International Journal of Cultural Property,* 29(1), pp. 63–80.

Gilchrist, R. (2020). *Sacred Heritage: Monastic Archaeology, Identities, Beliefs.* Cambridge: Cambridge University Press.

INAH (Instituto Nacional de Antropología e Historia) (2016). *Ética en la conservación del patrimonio cultural, mesa de discusión en la CNCPC*. Available at https://conservacion.inah.gob.mx/index.php/etica-en-la-conservacion-del-patrimonio-cultural-mesa-de-discusion-en-la-cncpc/.

INEGI (Instituto Nacional de Estadística y Geografía) (2020). *Censo de Población y Vivienda 2020.* Available at: www.inegi.org.mx/programas/ccpv/2020/

Isaac, B. L. (1983). Aztec Warfare: Goals and Battlefield Comportment, *Ethnology,* 22(2), pp. 121–131.

Knight, A. (2015). History, Heritage, and Revolution: Mexico, c .1910– c .1940, *Past & Present,* 226(s10), pp. 299–325.

Lira Vásquez, C. (2009). 'La Identidad "Colonial" De Oaxaca. Una Invención de la Política Turística y Patrimonial', in: A. Rodríguez Kuri and C. Lira Vásquez *(eds.) Ciudades Mexicanas Del Siglo XX: Siete Estudios Históricos.* Mexico City, Mexico: Colegio de México.

Lira Vásquez, C. (2014). El Discurso Patrimonial: Una Coartada Para Frenar El Desarrollo De La Ciudad De Oaxaca. *Revista Urbano*, May 2014. Available at https://revistas.ubiobio.cl/index.php/RU/article/view/238/205.

López, R.A. (2010). *Crafting Mexico: Intellectuals, Artisans and the State after the Revolution*. Durham, NC: Duke University Press.

Magaloni, B., Díaz-Cayeros, A. and Ruiz Euler, A. (2019). Public Good Provision and Traditional Governance in Indigenous Communities in Oaxaca, Mexico, *Comparative Political Studies,* 52(12), pp. 1841–1880.

Martínez Sola, María del Carmen (1998). *El Obispo Fray Bernardo de Alburquerque: El Marquesado del Valle de Oaxaca en el Siglo XVI.* Oaxaca: Instituto Oaxqueño de las Culturas; Fondo Estatal para la Cultura y las Artes; Secretaría de Desarrollo Turístico.

Meyer, J. (2007). 'Religious Conflict and Catholic Resistance in 1930s Oaxaca', in: M. Butler (ed.) *Faith and Impiety in Revolutionary Mexico.* Basingstoke: Palgrave Macmillan, pp. 185–202.

Morales, F. (1998). Mexican Society and the Franciscan Order in a Period of Transition, 1749-1859, *The Americas,* 54(3), pp. 323–356.

Mullen, R.J. (1975). *Dominican Architecture in Sixteenth-Century Oaxaca*. Tempe: Arizona State University.

Mundaca, A. (2023). 'Por no ser católicos, encarcelan a 4 personas en San Juan Lalana, pueblo chinanteco de Oaxaca.' *El Universal de Oaxaca*, 24 November, 2023. Available at: https://oaxaca.eluniversal.com.mx/municipios/por-no-ser-catolicos-encarcelan-4-personas-en-san-juan-lalana-pueblo-chinanteco-de-oaxaca.

Norget, K. (2006). *Days of Death, Days of Life: Ritual in the Popular Culture of Oaxaca*. New York: Columbia University Press.

Norget, K. (2008). Hard Habits to Baroque: Catholic Church and Popular-Indigenous Religious Dialogue in Oaxaca, Mexico, *Revista Canadiense de Estudios Hispánicos,* 33(1), pp. 131–158.

Norget, K. (2021). Popular-Indigenous Catholicism in Southern Mexico, *Religions,* 12(7), pp. 531.

Pérez Cruz, V., Torres Valdéz, R., Covarrubias, F. and Ojeda Aquino, E. (1991). Santa Cruz Mixtepec, Zimatlán Oaxaca, *Cuadernos de Arquitectura Virreinal,* 10, pp. 82–88.

Poole, D. (2004). An Image of 'Our Indian: Type Photographs and Racial Sentiments in Oaxaca, 1920-1940', *Hispanic American Historical Review,* 84(1), pp. 37–84.

Rodríguez, J. (2014). 'Oaxaca, patrimonio cultural de… Harp.' *Observatorio de Historia, 29 May, 2014.* Available at: https://elpresentedelpasado.com/2014/05/29/oaxaca-patrimonio-cultural-de-harp/

Rodriguez, O. (2021). Encarcelado y expulsado por no pagar la fiesta del Santo Patrono de municipio de Usos y Costumbres en Oaxaca, *Red Capital*, November 27, 2021. Available at: https://red-capital.com.mx/encarcelado-y-expulsado-por-no-pagar-la-fiesta-del-santo-patrono-de-municipio-de-usos-y-costumbres-en-oaxaca/.

Romero F. and de los Ángeles, M. (2000). 'The Indigenous Population of Oaxaca from the Sixteenth Century to the Present', in: R. Adams and M. MacLeod (eds.) *The Cambridge History of the Native Peoples of the Americas,* Vol.2, Part 2. Cambridge: Cambridge University Press, pp. 302–345

Rozental, S. (2016). In the Wake of Mexican Patrimonio: Material Ecologies in San Miguel Coatlinchan, *Anthropological Quarterly,* 89(1), pp. 181–219. DOI: https://doi.org/10.1353/anq.2016.0007

Salas Landa, M.M. (2024). *Visible Ruins: The Politics of Perception and the Legacies of Mexico's Revolution.* Austin: University of Texas Press.

Santa Cruz Mixtepec (2020). *Plan De Desarrollo Municipal Santa Cruz Mixtepec, Oax. 2020-2022*. Available at: http://sisplade.oaxaca.gob.mx/sisplade/smPublicacionesMunicipio.aspx?idMunicipio=378#divPlanesT

SECTUR (Secretaria de Turismo del Estado de Oaxaca). *Indicadores de la Actividad Turística*, (2018). Available at: www.oaxaca.gob.mx/sectur/informacion-de-interes/estadisticas/

Torres Torres, Fr. Eugenio Martín (1997). 'La Desolación de un Convento: Santo Domingo de México, 1860-1861', in: J.B.Barquilla and S.R. López (eds.) *Los Dominicos y el Nuevo Mundo, siglos XVIII-XIX: Actas del V° Congreso Internacional 1995*. Salamanca: Editorial San Esteban, pp. 309–348.

UNESCO (2023). UNESCO Database of National Cultural Heritage Laws. www.unesco.org/en/articles/unesco-database-national-cultural-heritage-laws-updated.

Valadez, R. (2024). 'Prevé Oaxaca aumento de 8% en llegada de turistas en 2024.' *Milenio*, 24 April, 2024. Available at: www.milenio.com/negocios/sectur-preve-aumento-llegada-turistas-oaxaca-2024.

Winter, M.C. (1978). *Cerro de la Cueva: Una zona arqueológica del clásico en Santa Cruz Mixtepec y San Bernardo Mixtepec, Distrito de Zimatlán, Oaxaca.* Estúdios de Antropologia e Historia, Numero 12. Oaxaca: Centro Regional de Oaxaca, Instituto Nacional de Antropologia e Historia.

Wood, W.W. (2008). *Made in Mexico: Zapotec Weavers and the Global Ethnic Art Market.* Indianapolis: University of Indiana Press.

13 Biocultural Heritage and Archaeology as Sources of Knowledge, Cultural Identity, Sustainable Tourism and Economic Resilience in Ligüiqui, Manabí, Ecuador

Juan Andrés Jijón-Porras, Manuel Castro-Priego, Fernando Represa-Pérez and Leonardo Alonzo-Zambrano

13.1 Introduction

Ligüiqui is an emblematic archaeological site in the province of Manabí, Ecuador, characterized by its maritime architecture, which is unique not only within the Pacific coast of Ecuador but also throughout the north-central region of South America (Figure 13.1). The site, located near Cape San Lorenzo (Manta), has a predominantly Manteña cultural base (*c.* 700 AD–1532 AD), although other coastal cultures belonging to the so-called Integration Period (500 AD–1532 AD) and Regional Development Period (500 BC–500 AD) have been documented. Ligüiqui has more than 1500 hectares of archaeological remains, whose main nucleus is located between rugged hills and the Pacific Ocean.

In the sea, the architecture is a large complex of predominantly semicircular 'marine pens', distributed in several sets along the coastline. These pens are associated with passive fishing methods of pre-Hispanic origin. Recent studies suggest a long occupation at Ligüiqui with an equally prolonged period of marine exploitation using the pens. The recovered evidence indicates a scale and diversity of fishing previously unreported in Ecuador. The people of Ligüiqui still use the marine pens to collect fish trapped after the low tide and cultivate short-cycle crops in the terrestrial area (Alonzo, 2015). As such, their fishing activity is an ancestral tradition that, at a minimum, would go back to the Manteña era.

This chapter describes recent archaeological research at Ligüiqui and how its cultural heritage might support the development of heritage tourism to enhance economic resilience of the community. Farming and fishing-based activities have been seriously disrupted in recent years by a succession of events: recurrent devastation brought by the El Niño phenomenon, the 2016 earthquake that ravaged Ecuador's coast and the COVID-19 pandemic. In addition, the community faces

DOI: 10.4324/9781003491071-16

Figure 13.1 Ligüiqui: geographic location. © Manuel Castro-Priego.

rising insecurity linked to drug trafficking, while the region has ongoing political and economic instability. Tourism to Ligüiqui has already undergone some development, but this has occurred without attention to the cultural heritage. We describe plans for a museum and how this may support not only tourism but also the protection of archaeological sites by enhancing community awareness about its own cultural heritage.

13.2 Archaeological Research in Ligüiqui

The importance of Ligüiqui's archaeological remains has long been recognized by its inhabitants and archaeologists familiar with the Manabí coast, with ongoing research documenting its unique character and cultural significance. Professional archaeological studies began in 2012 when the National Institute of Cultural Heritage of Ecuador (INPC) commissioned an archaeological prospection study with the purpose of creating a more systematic record of archaeological resources in the areas near the mouth of the Portoviejo River and that of Ligüiqui. This first study was directed by archaeologist Juan José Ortiz Aguilú, assisted by a team of Ecuadorian archaeologists and technicians and supported by members of the Ligüiqui commune. During that initial study, the existence of the marine pens was reported to the Ecuadorian scientific community, recognizing their importance and implications for the economy of the Manteño people at around 800 AD (Marcos, 2012; Ortiz, 2012).

Figure 13.2 Stone walls to retain terraces in the Los Charcos area, Ligüiqui. © Juan Andrés Jijón-Porras.

Subsequently, from 2013 to 2014, geoarchaeologist Cristian Favier joined the research team and was able to analyse the geomorphological characteristics of the site and their corresponding links to the human groups that settled there. It is worth noting that some of the community members of Ligüiqui who participated as assistants in the archaeological investigations continued their university studies related to topics linked to cultural heritage and community tourism, one of them publishing his thesis on the Manteños fishing corrals (Alonzo, 2015).

Archaeological research continued from 2014 to 2016 (Favier Dubois and Ortiz Aguilú, 2017; Favier Dubois et al., 2019; Figure 13.2). Since 2018, a collaboration between the University of Alcalá de Henares and the INPC (within the framework of the international cooperation project '*Perduraciones, Continuidad y Ruptura Nuevas realidades de desigualdad en la costa Ecuatoriana*' has defined the total extent of the marine pens in Ligüiqui. They extend for more than 10 km along the coast, and their methods of construction and resource extraction capacities have been examined (Olmo Enciso and Castro Priego, 2020). The coastal research was accompanied by an investigation of associated terrestrial archaeological remains, with the application of drone-based photogrammetry, magnetometry, excavation, ceramic analysis, extraction of paleoenvironmental and archaeofaunal samples and radiocarbon dating (Vásquez Sánchez et al., 2023; Jijón Porras et al., 2023; Jijón Porras, 2019).

13.3 The Settlement and Fisheries at Ligüiqui

Because of this research, the joint coastal and terrestrial site of Ligüiqui has become recognized as one of the most interesting complexes in Ecuador for the period of the Manteño culture (6th–16th c. AD). It represents an important addition to the pattern of pre-Hispanic settlement, joining Salango (Lunniss, 2019; 2020), Agua Blanca (McEwan, 2003), Cerro de Hojas-Jaboncillo (Marcos, 2012), Los Frailes (Mester, 1990) and López Viejo (Currie, 1995). Ligüiqui has several features that provide new information about the Manteño settlement organization. It is situated within a group of hills overlooking the Pacific Ocean and has occupational structures of a distinctive character, with parallels to those observed at the monumental Cerro de Hojas-Jaboncillo. At more than 2000 acres, the spatial extent of the terrestrial settlement is impressive, spread across hilly terrain with stone walls to retain terraces and associated with seasonal water channels that extend continuously along the coast (Figure 13.3).

Radiocarbon dating and the recovered material culture have established a chronology that extends throughout most of the Manteño period (6th–15th centuries), although earlier activity has been observed in specific areas, which date back to the Guangala period (500 BC to AD 550). The decline in occupation appears to occur in the mid-15th century, significantly before the European colonial presence. Although this decline remains to be further explored, it entailed a drastic reduction of the settlement, with Ligüiqui appearing to become what Spanish textual sources describe in the early 17th century as a small place, populated by only thirty-five inhabitants and on the verge of disappearing (Ponce Leyva, 1994, p. 35).

Figure 13.3 Marine and terrestrial archaeological research areas at Ligüiqui. © Manuel Castro-Priego.

The most important excavation has been carried out in the location known as Chácara I, where a large dwelling or tola is being explored, together with a group of agricultural terraces and warehouses in the immediate area. The central building measures 51x25 m and was built with masonry and adobe, on a platform that reaches a height of 4 metres. The excavation (454 m^2) has delimited what appears to be a residential complex, with one of the entrances in the form of a raised ramp, through a stone and adobe corridor. Domestic, kitchen, performance and cemetery areas have been documented. During the 2021 field season, two large rooms were located around the raised platform, which have been interpreted as storage areas. The chronology suggests that the complex had a long process of development, starting late in the 12th century, and was occupied until the mid-14th century AD.

One of the most interesting aspects is the close relationship between living spaces and agricultural production areas. Organic residues in the ceramic contexts show the importance of a diet structured around the consumption of maize. Agricultural production involved a system of water supply by means of large pits, some of enormous dimensions (47.50 m in diameter and a capacity of 20,000 m^3). These captured rainwater by using the humidity contained in the coastal fogs, especially during the dry season. Water storage structures from the Manteño period can be observed at the site of 'Los Charcos'.

The most unique element of Ligüiqui, however, is undoubtedly the extensive system of fisheries or marine pens, which had served as massive hunting grounds for fish and crustaceans. They were positioned around stone terraces or walls to form corrals, which extend into the Pacific Ocean for at least 300 m (Figure 13.4).

The marine pens were built from large, trapezoidal, slab-like blocks of basalt stone. Many were placed in a staggered arrangement, with the most regular trapezoidal side resting on the seabed. The height of the structures at low tide can reach 1.40 m, with walls or terraces between 1.30 and 1.88 m wide. These have two rows of slabs, the inner row being higher and with larger blocks than the outer row, in a sloping design, which prevents the removal of fish species at low tide and avoids the destructive action of the waves. The construction of each of the pens required excavating a two-metre-wide area on both sides of each of the walls, on which a bed of slabs was laid. The mortar of the walls was made of sand sludge with the addition of algae to form consolidated blocks, with basalt flagstones used

Figure 13.4 Marine corrals exposed at low tide in the southern area of Ligüiqui. © Juan Andrés Jijón-Porras.

for the base of the walls. The purpose of their construction was to trap fish and various types of shellfish; such activity can still be observed on the beach of 'Punta Cangrejo'.

All known corrals show a systematic organization, forming a network that suggests the capture of different species by those closest to the shore and those further away (Figure 13.4). The corrals closest to the shore were also used as a space for collecting shellfish and fishing for crustaceans and small fish by standing in the pen. Different methods of fishing likely occurred in those located further away from the beach. Three lines of pens have been documented at the southern end of Punta Cangrejo, although it is not yet possible to discern whether this is an 'organic' growth pattern, meaning the expansion of a design because of increased production, or was a single construction. A few pens had internal partitions using walls or small structures, although these are not necessarily found in the larger structure. In the Iberian Peninsula, these partitions are known as 'corralillos chicos' or 'atajos' (Florido del Corral, 2011). Their function is threefold: to enhance the accumulation of fish, to guide the process of water outflow and to disorient the fish so they find it difficult to escape.

In the case of the Ligüiqui pens, the structures have openings at the end of their upper vertices, which suggest a space that had once been blocked by vegetation or perhaps leather fragments. The pens form a semicircle. At the end that goes into the sea, there are openings that allow the fish to pass through. In the well-known corrals of Rota (Cádiz, Spain), the obstruction of the water outlet was made with a framework of vine remains, although nowadays a metal plate is used. In the case of Rota the origin of these structures is mediaeval, although they were used until the 20th century (Florido del Corral, 2011, Muñoz-Pérez et al., 2002).

The design of the Ligüiqui fisheries had required a deep knowledge of the environment, the available resources, the tidal cycles and the behaviour of the fish species, while their construction and maintenance had required strong community organization. This high level of investment enabled a very effective passive fishing system. The Ligüiqui fisheries have some similarities to those of Chiloé (Chile) (Álvarez et al., 2008), where the weight of fish caught per capture can exceed one tonne. In addition to the passive fishing, the use of rafts, canoes and fishing nets is indicated by the finds of punches and weights made of basalt and granite with a central recess, from our excavations.

Regarding the fish species caught, an initial identification was made of the ichthyofauna assemblages recovered from the excavations, which account for 90% of the fauna recovered. Among the species, the following stand out: jack mackerel (*Trachurus simmetricus*), sierra (*Thyrsite atun*) and hake (*Merluccius gayi*), together with other species such as Conger (*Genypterus sp.*), Corvina (*Cilus gilberti*), Rolizo (*Pinguipes chilensis*) and snook (*Eleginops maclovinus*) (Álvarez et al., 2008, p. 114), parrotfish (*Scarus*). The ichthyofauna of Ligüiqui shows a more local character than that from nearby settlements such as Salango (Beárez et al., 2012) (1,500–100 B.C.). In the latter, tuna and other pelagic fish predominate, species also observed in Ligüiqui but at much lower intensities.

The malacological assemblage from the Ligüiqui site presents a wide diversity of species, with 50 species of marine molluscs and one genus-level taxon of a terrestrial gastropod. No freshwater gastropods have been recorded (Vásquez Sánchez et al., 2023). All the species correspond to the Panamic Malacological Province, with some known to have a wide distribution (*Thais haemastoma* and *Thais melones*). According to Keen (1971), four species have a narrow geographical distribution: *Tegula picta*, *Polinices navidus*, *Thais delessertiana* and *Protothaca zorritensis*. However, the ancient geographical distributions were likely different from the present ones, while molluscs might have been acquired by trade with communities to the north and south of Ligüiqui.

As a result of the finds at Ligüiqui, similar fishing sites are beginning to be discovered elsewhere, such as those at Cabo Pasado (San Vicente, Manabí). All of them may have been used both in pre-Hispanic times and in the later colonial period. However, the absence of references in Spanish textual sources from the 16th and 17th might indicate the disappearance of the fisheries at that time, coinciding with the introduction of new fishing techniques and exploitation of the environment.

13.4 Community and Heritage Tourism in Ligüiqui

Ligüiqui has excellent prospects for the development of tourism based on the archaeological–historical heritage we have described. This could provide a significant stream of income for the local community at a time when both fishing and agricultural economies have been, and continue to be, challenged by environmental events, long-term climate change and political instability. While the cultural heritage of Ligüiqui provides an opportunity for tourist initiatives (Alcivar and Noor 2019), it also needs protection in the face of pressure from external agents such as the real estate developers and drug traffickers (Jijón and Rivera, 2020). Drug trafficking has plagued many Latin American countries for decades, and in recent years it has taken its toll on Ecuador. Specifically in the Manta area, this has had devastating impacts, including the murder of its Mayor Agustin Intriago on July 23, 2023. This shocked the entire population and has led to an environment of permanent tension.

Drug trafficking has an impact on the practice of archaeology and development of tourism by generating a climate of fear and mistrust in local communities, which makes collaboration with researchers difficult and can lead to the loss of valuable information. However, this still does not represent a total impediment to the development of archaeological research and tourism thanks to the sustained efforts to consolidate the Ligüiqui as a point of high archaeological interest at the national level.

On the positive side, the Ligüiqui archaeological sites and natural environment are already frequently visited, the area being used for experiential fishing, sport fishing and recreational activities. These activities lead to a key question: How can archaeological structures be preserved, protected and exhibited at the same time?

Archaeo-tourism, also known as heritage tourism, is a form of tourism that encompasses locations, activities and services in which archaeology is the main ingredient and the motivation for those undertaking a trip, whether it be from short, medium or long distances and of differing durations (Tresserras, 2004; Manzato and Rejowki, 2007). This definition of heritage tourism can also encompass the process of discovery, movement or stay of visitors to so-called archaeological sites, where the vestiges of ancient societies, whether historical or prehistoric, are found and can be visited by land or water.

Through the research and studies that have been carried out at Ligüiqui, it is intended to identify the opportunities and limitations to optimize heritage tourism as a source of income. The development of such tourism has already been supported by the Manta Municipality, the Manabí Provincial Government, the National Institute of Cultural Heritage and the Ministry of Environment and Water. With the dissemination of information about Ligüiqui's natural and cultural attractions, it has been possible to stimulate tourism activity by improving hotel infrastructure, restaurants, recreation, transportation services and handicrafts. The new income streams linked to tourism should benefit the local inhabitants, by improving their quality of life and deterring them from migrating to the cities, where in most cases they will be forced into poverty.

It is important to provide information about each of the different archaeological sites in Ligüiqui, not only for the people who inhabit this territory but also for those who come to know and learn more about Ligüiqui's cultural heritage. As such, the 'Underwater and Land Archaeological Route' project was launched. This initiative is the result of academic research by the then President of the Ancestral Commune of Ligüiqui, who together with his council launched an entire socioeconomic revitalization strategy through heritage tourism. The primary objective was to promote cultural, scientific, educational development and sustainable tourism for the 400 inhabitants of Ligüiqui and the surrounding communities of the San Lorenzo parish and the Manta canton. For this purpose, tourist trails were defined in the terrestrial archaeological areas, connecting different points of interest including the system of marine corrals that extend along the coast. In that sense, a circuit has been designed that allows immersion within the archaeological sites, with an interpretation centre. The route is in its first phase of implementation thanks to the support of several institutions, but notably the efforts of the community and its dedication to recover its cultural heritage.

That's how the 'Young Guardians Group' was created, with the aim of safeguarding terrestrial and marine archaeological spaces (Figure 13.5). This project seeks to inform young people and adults about the archaeological sites and encourage them to further disseminate this information (Alonzo, 2022). It welcomed thirty-three young volunteers from the Ligüiqui community, who undertook cleaning activities on the beaches and reforestation in the forests. The young people organized themselves to watch over the heritage site and actively supported scientific research. The group participated actively and on a rotating basis in archaeological investigations. They were part of the necessary staff supporting the clearing of vegetation and cleaning areas of interest. They assisted in the topographic

Figure 13.5 Participation of the group of young guardians of Ligüiqui. (a) Intervention in archaeological areas. (b) and (c) Cleaning of vegetation in the Los Charcos area. (d) Intervention in marine pens. (e) Assistance to archaeological survey with magnetometer. © Juan Andrés Jijón-Porras and Leonardo Alonzo-Zambrano.

survey and the management of the total station and were inducted into the use of drones. They were trained as technical assistants in archaeological surveys and excavations, processing material such as pottery, with permanent supervision and training by the professional archaeologists. The group also participated in cultural mediation with visitors as part of the programme for future tourist guides, as well as in night activities of interpretation of the marine corrals and experiential fishing.

13.5 From Heritage to Museum as a Strategy for Resilience and Sustainable Development

In view of Ligüiqui's outstanding importance, regarding both nature and culture, especially as it relates to pre-Columbian historical evolution, several museum proposals have been put forward during the last decade. One or more museums could play an important role in the preservation of Ligüiqui's cultural heritage and ensure its value is communicated to future generations. By supporting heritage tourism, a museum would increase the economic resilience of the community by countering the sharp decline in fishing that has created a crisis in Ligüiqui's traditional subsistence model.

The creation of a museum would not only support sustainable tourism and development but also help restore the links between the local community with their past that, as we have found in our fieldwork, have been largely lost. To date, none of the museum proposals have been implemented, either because of their technical complexity, high budgets or due to other institutional conditions specific to the country, which are accentuated in the cultural field. The current proposal is found within the World Bank's Manta 2035 Plan, which seeks to develop the port city of Manta, in Manabí Province, into one of Ecuador's most modern cities. The plan involves a museum archaeological park in Liguiqui for the conservation and enhancement of biocultural heritage (Decentralized Autonomous Government of the Manta Canton, 2024).

The museum proposal combines classic elements of an eco-museum (territory, heritage and community) with features of recent interpretative centres such as citizen participation and the incorporation of digital technologies.[1] This proposal finds a local precedent at the southern end of the province of Manabí in the proposal of a 'ecomuseum of the fishing communities of the Puerto López canton', conceived as "a participatory space for protection, socioeconomic dynamization and empowerment of the communities based on the enhancement of their biocultural heritage" (Represa et al., 2023, p. 220). In this case, the intention is not to construct a costly building that would be difficult to maintain and would eventually lead to its abandonment (a 'white elephant' according to popular Ecuadorian terminology), but rather to generate a multifunctional space, associated with itineraries distributed throughout the territory. That makes it possible to reconcile the functions of conservation and enhancement, influencing the latter through awareness and citizen participation activities that contribute to developing a community debate about its identity (always subject to negotiation) and the importance of maintaining its collective memory.

Within this conception, we find it interesting to highlight the experience of the Finca 6 Museum Site Visitors' Center in Diquís Celta (Costa Rica). This is part of the first set of archaeological assets declared as World Heritage Sites by UNESCO in Costa Rica. The visitor centre combines the conservation of archaeological, historical, cultural and environmental heritage with educational functions aimed at the communities (Morales, 2023).

Knowledge and heritage awareness are two fundamental principles of the Finca 6 Museum Site initiative that has been operating successfully for more than a decade. Its location and context have parallels with Ligüiqui: both are located in a marine-coastal territory on the shores of the Pacific Ocean, share aspects of their colonial historical development and have had their archaeological heritage recently discovered. Moreover, the most impressive monuments in both locations are of stone. In the Costa Rica case, these are stone spheres of a surprising size and in Ligüiqui the use of stone to build the fisheries. The Ligüiqui community is currently changing its perception of those stones. They are becoming recognized as a testimony to an indigenous past, of a complex society with the capacity to develop specialized architectures reflecting a high level of economic, political and social organization.

Enhancing this change of perception will be key for both preserving the stone structures and enabling the development of heritage tourism. In this social process of re-signification of objects from the past, a museum can play an important role, providing it maintains a strong relationship with the community. This is the intention, with current planning involving meetings with the community, so the final design is not an imposition but expresses and channels their values and concerns.

Overall, the aim is to avoid the notion of an emblematic building for the display and interpretation of archaeological remains with a more holistic proposal that integrates a dialogue of feelings and knowledge between academics and the inhabitants of Ligüiqui. The museum will project both towards the marine-coastal territory with its fishing facilities and inland to the terrestrial settlement. It will incorporate an educational/sociocultural aspect oriented towards the mobilization of community commitment to its past but also to its future through the defence of its biocultural heritage. By promoting community empowerment and self-management, enhancements in gender equality and sustainability will also be achieved. The holding of workshops and sociocultural activities will have a special relevance in the implementation of the proposal, exploring how to combine the enhancement of biocultural heritage, the support of scientific research and enhanced cultural and environmental awareness.

The promotion of community and sustainable tourism will receive special attention within these workshops. They will aim to embed the role of tourism as part of building economic resilience for the community, as has been achieved in other coastal areas of Latin America (González Herrera, 2022). Finally, a further level of implementation of the museum project in Ligüiqui contemplates the generation of collaboration networks with other museums and protected sites at a local and international level.

13.6 Reflections

Archaeological research in Ligüiqui, together with the social and economic conditions of its current inhabitants, has allowed the development of a vision in which the future safeguarding of biocultural heritage is at the centre of proposals that link cultural identity with sustainable socioeconomic development. However, there remain significant risks arising from the pressure of actors external to the community: urban planning that is opposed to the protection of cultural heritage and ecosystems, the growth of social and political instability linked to drug trafficking and the need to consolidate research projects that will ensure the conservation of archaeological sites.

In this sense, the case of Ligüiqui is presented as a case study that reflects the reality of the link between cultural heritage, society and its challenges in development not only in this place, but also in multiple archaeological sites in the region and the country. Ecuadorian archaeology is very rich, with immense potential for scientific research, socio-cultural linkage and tourism promotion. Unfortunately, Ecuadorian archaeology remains largely unexplored and is still invisible as a driving force for development for the national and international public, with few significant initiatives in recent years and without major prioritization by institutions. In this context, Ligüiqui still has time to embark on a process that allows the preservation of cultural heritage with the modernization of the economy, provided this process is linked to research, education and the commitment of its guardians.

Acknowledgements

We would like to first thank the Ancestral Community of Ligüiqui for opening the gates of their homes, families and memories to us. We are also grateful to the public and academic institutions that have provided fundamental support for the research: ArqueoSapiens S.A., INPC, Laica Eloy Alfaro University, University of Alcalá de Henares, Pacoche Marine and Coastal Wildlife Refuge and Municipality of Manta.

Note

1 In 2022, ICOM (International Council of Museums) approved a new definition of a museum that is in tune with the transition challenges that we have discussed, focusing on service to society, inclusion and citizen participation, as well as the importance of considering aspects such as diversity and sustainability.

References

Alcivar, L. and Noor, F.N.M. (2019). *Ligüiqui, potencial destino de turismo comunitario arqueológico en Manta Ecuador*, TURYDES: Revista sobre Turismo y Desarrollo local sostenible, ISSN-e 1988-5261, 12(27), pp. 1–18.

Alonzo, L. (2015). *Estudio de los corrales marinos en las zonas arqueológicas de la comuna Liguïqui, Cantón Manta, como un atractivo turístico potencial para la comunidad.* Manabí: Master Thesis in Tourism, Universidad Laica Eloy Alfaro.

Alonzo, L. (2022). *Estudio de los Corrales Marinos de la Comuna como aporte al desarrollo turístico.* Master Tesis, Universidad Laica Eloy Alfaro.

Álvarez, R., Munita, D., Fredes, J. and Mera, R. (2008). *Corrales de pesca en Chiloé.* Santiago de Chile: Consejo Nacional de Cultura y las Artes.

Béarez, P., Gay, P. and Lunniss, R. (2012). Sea Fishing at Salango (Manabí Province, Ecuador) during the Middle Formative Machalilla Phase, *Latin American Antiquity*, 23(2), pp. 195–215.

Castro-Priego, M., Olmo-Enciso, L., Labrada-Ochoa, M., Jijón-Porras, J., and García-Campoverde, J. (2021). Agrarian Spaces, Pre-Hispanic Settlement and LiDAR Technology in the Central Coast of Ecuador, *Virtual Archaeology Review*, 12(25), pp. 140–157.

Coral Almeida, B. and Represa Pérez, F. (Comps.): *Natura, Cultura y Desarrollo: Retos en Perspectiva Latinoamericana.* Manta: Editorial Mar Abierto.

Currie E. (1995). *Prehistory of the Southern Manabí Coast, Ecuador.* Oxford: British Archaeological Reports International Series 618.

De Sousa Santos, B. (2011): Epistemologías del sur. *Utopía y praxis latinoamericana*, 16(54), pp. 17–39.

Escobar, A. (2021). Reframing Civilization (s): From Critique to Transitions. *Globalizations*, pp. 1–18. DOI: https://doi.org/10.1080/14747731.2021.2002673

Favier Dubois, C. and Ortiz Aguilú, J. (2017). 'Geoarqueología en la costa manabita: Cambios en el paisaje costero y su importancia para las ocupaciones humanas de ayer y hoy', in: M.F.

Ugalde (ed.) *Volcanes, cenizas y ocupaciones antiguas en perspectiva geoarqueológica en América Latina*. Quito: Estudios de Antropología y Arqueología 2, Ediciones Pontificia Universidad Católica del Ecuador, pp. 115–128.

Favier Dubois, C., Storchi Lobos, D., Lunniss, R., Mora Mendoza, A. and Ortiz Aguilú, J. (2019). Pre Hispanic Fishing Structures Preserved on the Central Coast of Ecuador, *Journal of Maritime Archaeology,* 14, pp. 107–126.

Florido del Corral, D. (2011). 'Los corrales, una técnica de pesca tradicional en Andalucía', in: D. Bernal (ed.) *Pescar con arte. Fenicios y Romanos en el origen de los aparejos andaluces*. Cádiz: Universidad de Cádiz, pp. 65–94.

Gobierno Autónomo Descentralizado del Cantón Manta (2024): Plan de Desarrollo y Ordenamiento Territorial (del Cantón Manta (PDOT): Plan Manta Infinita 2035. Alcaldía de Manta.

González Herrera, M.R. (2022). 'El aprendizaje turístico como iniciativa para la construcción de resiliencia en comunidades pesqueras de América Latina', in: F. Represa Pérez (ed.) *Territorios Pesqueros: Resiliencia, saberes locales y cambio en Latinoamérica*. Manta: Editorial Cuerpo de Voces, pp.147–200.

ICOM [International Council of Museums] (2023) Museum Definition. Available at: https://icom.museum/en/resources/standards-guidelines/museum-definition/ [Accessed: 19/04/2024].

Jijón Porras, J. (2019). Arqueología subacuática en el Ecuador: Estado de la cuestión, Revista de Historia, Patrimonio, *Arqueología y Antropología Americana*, 1, pp. 5–21.

Jijón Porras, J., Pacheco Navarro, C. and Labrada Ochoa, M. (2023). 'Pre-Hispanic Subsistence Strategies on the Central and Southern Coast of Ecuador: Ligüiqui and Puná Island', in: D. Elkin and C. Delaere (eds.) *Underwater and Coastal Archaeology in Latin America*. Gainesville: University Press of Florida.

Jijón Porras, J. and Rivera, A. (2020). Arqueo-turismo en Manabí: activación turística del legado ancestral para el desarrollo local en la provincia de Manabí, Ecuador, Revista de Historia, Patrimonio, *Arqueología y Antropología Americana*, 3, pp. 75–90.

Keen A.M. (1971). *Seashells of tropical west America: Marine molluscs from Baja California to Peru.* Stanford, California: Stanford University Press.

Khotari, A., Salleh, A., Escobar, A., Demaria, F. and Acosta, A. (2019). *Pluriverse: A Post-Development Dictionary*. Nova Delhi: Tulika Books, Authorsupfront.

Leff, E. (2012). Pensamiento ambiental latinoamericano: patrimonio de un saber para la sustentabilidad, *Environmental Ethics,* 34 (Supplement), pp. 97–112.

Lunniss, R. (2019). 'Huaca Salango. A sacred center on the Coast of Ecuador', in: C. Lozada and H. Tantalean (eds.) *Andean Ontologies*, New Archaeological Perspectives. Gainesville: University Press of Florida, pp. 49–78.

Lunniss, R. (2020). 'Manteño, Geography and Culture of', in: C. Smith (ed.) *Encyclopedia of Global Archaeology*. Springer.

Manzato, F., and Rejowki, M. (2007). *Evaluación del potencial turístico de sitios arqueológicos*. Estudios y Perspectivas en Turismo, 16(1), pp.72–91, Buenos Aires: Centro de Investigaciones y Estudios Turísticos.

Marcos, J.G. (2012). 'La Investigación Arqueológica', in: J.G.Marcos, CI. Veintimilla Bustamante, O. Tobar, R. Lunniss and M. Suárez-Capello (eds.) *La sociedad prehispánica manteña en los Cerros Hojas-Jaboncillo.* Montecristi: Corporación Ciudad Alfaro. Boletín Arqueológico No.1, pp. 27–47.

McEwan, C. (2003). *And the Sun Sits in His Seat: Creating Social Order in Andean Culture.* Ph.D. dissertation, University of Illinois at Urbana-Champaign.

Mester, A.M. (1990). *The Pearl Divers of Los Frailes: Archaeological and Ethnohistorical Explorations of the Sumptuary Goods Trade and Cosmology in the North and Central Andes*. Urbana-Champaign: Ph.D. dissertation, University of Illinois.

Muñoz-Pérez, J.J., Acha Martín, A. and Fages Antiñolo, L. (2002). Los corrales de pesca en la costa Gaditana: siglos de entender el mar y sus recursos, *Revista de Obras Públicas*, 3(428), pp. 51–57.

Olmo Enciso, L. and Castro Priego, M. (2020). *La Cultura Manteña de Ecuador*, National Geographic. Available at: https://historia.nationalgeographic.com.es/a/cultura-mantena-ecuador_15489

Ortiz Aguilú, J. (2012). *Informe de la prospección y reconocimiento arqueológico de las áreas de Liguïqui y La Elvira*. Portoviejo: Technical report presented to the Instituto Nacional de Patrimonio Cultural del Ecuador.

Ponce Leyva P. (1994). *Relaciones Histórico-Geográficas de la Audiencia de Quito* (s. XVI-XIX). Quito: Ed. Abya-Yala.

Represa, F. and Vina, M. (2021). From Below and From Within: Fishing Communities Under the COVID-19 Pandemic and Other Globalizations in Southern Manabi, Ecuador. In: *Territory, Politics, Governance*. DOI: https://doi.org/10.1080/21622671.2021.1960595

Represa Pérez, F. (2023). 'Introducción. Transiciones bioculturales. Desafíos y propuestas locales', in: F. Represa Pérez (ed.) *Transiciones Bioculturales en territorios marino-costeros de Latinoamérica* Quito. Ecuador: Ediciones Mawil, pp. 16–20.

Represa Pérez, F., Font Aranda, M., Zambrano Varela, B. and Hidalgo Santos, L. (2023). 'Ecomuseo de las Comunidades Pesqueras del Cantón Puerto López: Diseño de recorrido interpretativo', in: F. Represa Pérez (ed.) *Transiciones Bioculturales en territorios marino-costeros de Latinoamérica*. Quito. Ecuador: Ediciones Mawil. pp. 220–225.

Ribeiro, G.L. (2009). Non-Hegemonic Globalizations: Alter-native Transnational Processes and Agents, *Anthropological Theory*, 9(3) pp. 297–329.

Tresserras, J. (2017). *El conjunto conventual de San Francisco de Quito: un caso piloto de programa de patrimonio mundial y turismo sostenible. Los retos de la gestión turística responsable y la participación comunitaria*, Anales de la Universidad Central del Ecuador, N°373. Quito: Universidad Central del Ecuador. pp. 511–531. DOI: https://doi.org/10.29166/anales.v1i373.1363

Vásquez Sánchez, V. F., Rosales Tham, T., Castro Priego, M., Olmo Enciso, L., Jijón Porras, J.A., Labrada Ochoa, M.O. and Morales Muñiz, A. (2023). 'Subsistence intensification: the molluscs from the Chácara I sector at Ligüiqui (Manabí Province, Ecuador: AD 12th-15th centuries)', in: N. Pöllath, N. Battermann, V. Goebel, S. Emra, P.D. Paxinos, M. Schwarzenberger, S. Trixl and M. Zimmermann (eds.) *Animals and Humans through Time and Space: Investigating Diverse Relationships*. Essays in Honour of Joris Peters. Rahden VML Verlag Marie Leidorf, pp. 407–425.

14 The PIARA Peru Project at Hualcayán

An Archaeology and Partnership of Care

Rebecca Bria and Erick Casanova Vásquez

14.1 Introduction

The PIARA Peru project is a community-centred research project in the highland Andes of Huaylas, Ancash, Peru. PIARA, which stands for the *Proyecto de Investigación Arqueológico Regional Ancash*, was established by Rebecca Bria in 2009 as a dissertation research project with a foundational goal to invest long-term in the communities where our archaeological work was undertaken. By 2010, our research was firmly established at the archaeological site and in the community of Hualcayán, a rural *comunidad campesina* or communal land holding 'peasant' community (Figure 14.1). Our mission was threefold: (1) to conduct archaeological investigations that reveal the region's Indigenous history; (2) to co-create research, heritage, and development projects with the Quechua and Spanish-speaking communities whose past we investigate; and (3) to educate the next generation of archaeologists through field schools and research opportunities.

Here, we reflect on our journey over the past 15 years working with the community of Hualcayán and in partnership with the regional university, the Universidad Nacional Santiago Antúnez de Mayolo (UNASAM). Underscoring our mission is an effort to cultivate a relationship of mutual care as we maintain a long-term presence in Hualcayán and the region more broadly.

Early on, the PIARA project's leadership[1] identified the importance of designing a research project that positively impacted the communities we work in. A top priority was to avoid an extractivist research design, such as one that prioritizes research questions over people, takes data and archaeological materials out of communities and never returns them, or does not provide any substantive benefit to local people. Hualcayán's densest archaeological remains lie near the centre of the contemporary village, making nearly all PIARA project decisions, activities, and preservation efforts regarding ancient remains consequential to the lives of local people (Figures 14.1 and 14.2).

Located on the eastern side of the Callejón de Huaylas valley, both the contemporary village and ancient site lie at the juncture of a sloping plateau and steep mountainside near 3150 metres above sea level (m asl). Hualcayán's archaeological remains are vast, with ancient religious, domestic, and mortuary features clustered within a 150-hectare area and extensive agricultural fields, terraces, and

DOI: 10.4324/9781003491071-17

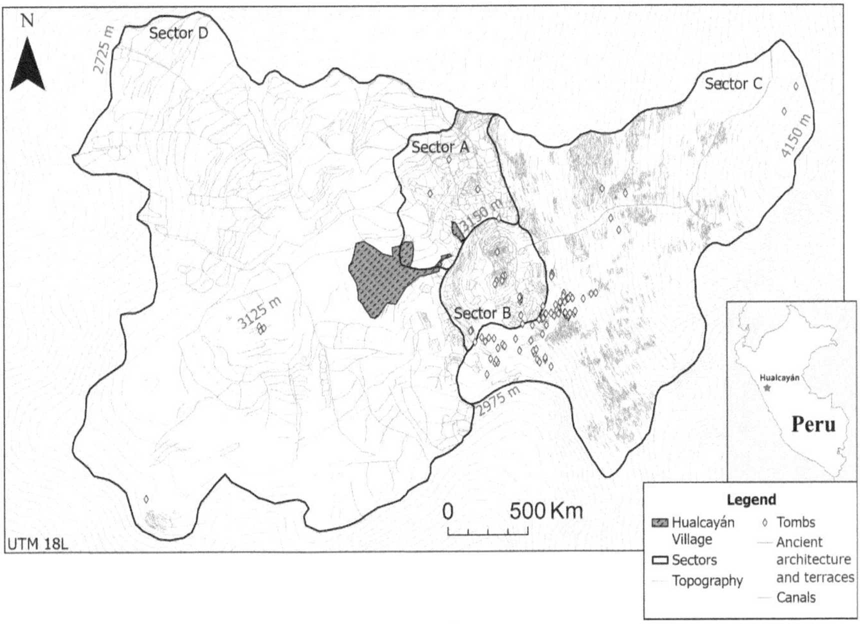

Figure 14.1 Map of Hualcayán showing the spatial relationship between the contemporary village and its extensive archaeological remains. These ancient remains are concentrated in sectors A and B, but ancient agricultural features and tombs extend throughout sectors C and D. Many families have a secondary home within their agricultural fields located beyond the core village area, where they reuse ancient infrastructure, such as terraces, canals, and reservoirs. © Rebecca Bria.

canals with interspersed tombs spreading to the east and west between 2700 and 3900 m asl. The Callejón de Huaylas is a long intermontane valley made up of two parallel ranges, the Cordillera Blanca and the Cordillera Negra, and Hualcayán is located on the western slopes of the Cordillera Blanca where towering glaciers have fed irrigation canals for millennia. The rhythm of life is largely regulated by farming activities that are tailored to the seasons, which divide the year into a rainy summer from October to April and a dry winter between May and September. Until recently, local farmers grew a typical variety of highland Andean and colonial-introduced crops, including maize, potato, quinoa, amaranth, beans, wheat, and barley, which they sold in local and regional markets. During the last decade, however, many farmers have shifted to producing crops that are popular for export, especially sugar snap and snow peas or strawberries (see Bria and Walter, 2019). Most households own a small variety of animals for consumption, including pigs, chickens, guinea pigs, sheep, and goats, while steers, horses, and donkeys are used as draft animals or for other field and transport tasks. Horses and donkeys, along with the paid expertise of their owners working as muleteers (*arrieros*),

Figure 14.2 Archaeological remains and agricultural fields are juxtaposed throughout Hualcayán. Here, the retaining walls and surfaces of two early religious mounds that were built during the Andean Formative Period (circa 2400–1 BC) are repurposed as field shelters (top) and agricultural terraces (bottom). © Rebecca Bria.

are also rented to Peruvian and foreign visitors to assist on trekking expeditions disembarking from Hualcayán; these visitors are typically brought to Hualcayán by trekking companies from the nearby cities of Caraz and Huaraz.

Hualcayán's extensive archaeological remains reveal a large ancient settlement with a long, complex history that spanned nearly 4000 years. This started around 2400 BCE with the construction of an early temple and extended to the time of the Inka Empire; it is likely that people were forcibly moved from Hualcayán by the Spanish soon after their arrival in 1532 CE. Hualcayán's prehistory is comparatively rich, bearing the remains of many successive societies with distinct material cultures (Bria, 2017). When establishing the PIARA project at Hualcayán, the first season of survey and test excavation indicated that ancient Hualcayán was an exceptional place to explore how a community can change dramatically over four millennia within a single village landscape. It also provided an opportunity to study a wide variety of research questions about the local past – be they motivated by the community or the archaeological team – and to partner with a rural community that by many accounts was fairly marginalized, even within its small district. At the start of the project, the community had an estimated population of just 300–400 people, while closer to 800 people live there today. Finally, we recognized that locked within Hualcayán's ruins was a vast but undervalued Indigenous history, suppressed by colonial violence and its complex legacy, that had included political efforts to 'modernize' the countryside in ways that sought to minimize its Indigenous identity (Ccahuana Córdova, 2020) and movements within Andean communities to adopt *mestizo* identities in the face of deeply entrenched racism (de la Cadena, 2000).

14.2 Project Funding, Design, and Development

For the first half of its existence, PIARA was a Ph.D. dissertation project. As such, it needed to ensure that both data collection and community-centred activities could be achieved within a relatively limited budget. The early budget combined doctoral dissertation grants from the National Science Foundation, the Wenner Gren Foundation, the Fulbright-Hays/IIE Graduate Fellowship for International Study, the Philanthropic Educational Organization, and various research and outreach grants from Vanderbilt University, while later faculty grants have come from the University of Texas at San Antonio, the Brennan Foundation, and the National Endowment for the Humanities, with more grants currently under review to expand the project. In the years 2011–2013, 2018–2019, and 2024, PIARA also ran an archaeological field school, which augmented the project budget.

We cautiously incorporated student learning into the PIARA project's mission, insisting that the field school could not be something that was uncritically tacked onto the project because we recognized that groups of foreign students would dramatically transform the way we interacted with the community and inhabited space within the village. We were transparent about how the PIARA field school fees would support some of our community-engaged activities, which were typically performed with the students' direct involvement as part of their learning. In fact, many students who chose the PIARA field school did so because our training included participation in community-focused research practices or service-learning activities. Attending students were also made aware that a portion of their fees would support several Peruvian university students from UNASAM, a university located in the city of Huaraz approximately three hours from Hualcayán, to join the field school at no cost.

The UNASAM student participation helped balance a strong foreign (principally North American and English-speaking) presence on the project and ensured that it was providing benefit to the local archaeology community as much as it was to the community where the research was performed. The UNASAM student participation was formalized with a *convenio*, or formal partnership agreement, between PIARA and the UNASAM School of Archaeology. The partnership stipulated that in exchange for course credits from the university, the School of Archaeology would receive desired equipment each year the field school was active, from total stations to GPS units, and that selected students would attend free of charge. UNASAM's School of Archaeology was fairly early in its development, and we saw our *convenio* as mutually beneficial to both the PIARA project and UNASAM students. Per UNASAM's request, each year the *convenio* is active we create an addendum to update the terms of the *convenio* for that year.

It quickly became clear that a long-term project required physical space to maintain a sustained presence in the community, to provide space for students and laboratory work, and to ensure that collected archaeological materials remained in the village rather than being permanently moved to a storage facility in the regional museum three hours away. Early on, the research team spent a few seasons of fieldwork living in the community's 'main building,' which we helped improve by building two bathrooms and showers; during the post-season lab work, we rented a house in the nearby city of Caraz.

In 2011, we created a small lab space in Hualcayán by constructing a large one-room structure on a land parcel next to the house of one community member who offered it to us when we were looking for somewhere to build. Later that year, two initiatives requiring community space, namely government approval for a small nurses' clinic and the arrival of a European agricultural company interested in working with local farmers, made it impossible for our research team to remain in the community space without impinging on other, more urgent community needs. Thus, in 2012 the PIARA team decided to build a more permanent structure. The community was able to offer land for its construction, but this was in an area with no running water and no immediate plans for its provision. While generous, it would have been exceedingly difficult to do our work and host students while maintaining sanitary conditions in a house without running water, so we weighed different options. At this moment the same community member who offered his land to build our initial laboratory space offered a second plot of land on his parcel to build a permanent lab and dormitory. In the same year, we completed the construction of a rustic 460 m^2 adobe structure in the village that housed a laboratory, artefact storage, and living quarters for up to 25 people (Figure 14.3).

Figure 14.3 View of Hualcayán with the PIARA lab and dormitory house visible in the lower right. © Rebecca Bria.

The house's construction on land assigned to one community member within a communal landholding community has proved contentious, but the building's existence was also essential to the project's success, especially for our ability to keep all artefacts within the community, and thus build trust that we were there to learn, not extract. This has been especially key because the community talks about collectors who had long come to Hualcayán to loot ceramic, textile, and metal objects from the site's hundreds of above-ground tombs.

14.3 Generating Relationships of Care through Collaborative Projects and Partnerships

As presented in the compelling essays within the recent volume *Archaeologies of the Heart* (Supernant et al., 2020), relationships of care between archaeologists and communities require a 'heart-centred' approach that foregrounds respect and understanding in all interactions and decisions, that operates with humility and openness, and that honours the lives of both the living community and the ancient and historic people who are the subjects of our research. PIARA has long operated according to these principles, with inevitable missteps corrected along the way as we learn from the people of Hualcayán, specifically what constitutes respect, support, and care within their community. Humility and openness have been key to this: we accept when we are wrong and respect what our community partners value, including when it does not align with what we value as a team of investigators. In the simplest terms, we place our effort in practicing an 'archaeology of care' (Caraher, 2019), whereby care is the guiding mode of engagement in all aspects of our archaeological project, from the manner in which we engage with ancient and historic things to how we honour commitments to and build relationships with the people where we work.

Actions that cultivate a relationship of care in Hualcayán have taken many forms and are constantly changing. On the one hand, our heart-centred work often has little to do with archaeology or heritage and more with responding to requests to assist on a community problem, such as when they fall short of the cost to complete the construction of a new schoolroom, build a library for their school, or grade the dirt road after heavy rains. Like many archaeologists who work in low- and middle-income communities, we also respond to individuals who knock on our door seeking support in times of personal crisis. Because this can lead to discontent between community members, we try to find ways to support individuals through work if they are able, or through community initiatives that have a broader reach.

In return, the community shows support for our project in big and small ways, including when they invite us back year after year, when they respectfully consider our ideas for research and collaboration, and when they greet our students on the street to make them feel welcome in their community. Supporting community projects through financial means or direct participation is not only to demonstrate care and build trust, however. In an Andean community, particularly a *comunidad campesina* like Hualcayán where membership and land rights are constituted through participation and/or redistributive contributions vis-à-vis work parties, labour rotations, and household collections. Therefore, they are essential social acts.

On the other hand, we have spent considerable time collaborating on projects that centre archaeology or heritage more broadly. Given that archaeology is of interest but not a top concern for the people of Hualcayán – having a strong harvest and providing for basic family needs rightfully take priority – these initiatives are co-designed to ensure that our projects are relevant to local people. As we have written about elsewhere, over the years we have moved towards a co-creative model for community projects, which collectively determine the direction and use of our project funds or efforts and have resulted in assembling a small community museum, organizing a festival, and beginning a women's handicraft enterprise (Bria and Cruzado Carranza, 2015; Connolly et al., 2015).

The heritage festival included contemporary dance, music, and games organized by the community with excavation tours, pop-up museum exhibits, and archaeology talks organized by the archaeological team to ensure the event was a celebration of the community today as much as a showcase of the ancient Hualcayán community (Figure 14.4). More recently, we have answered the call by science

Figure 14.4 Co-created events and projects in Hualcayán combine the knowledge, value, and interests of the archaeological team and the community. Pictured are flute and drum musicians and dancers leading a procession to the Hualcayán excavations during a co-sponsored heritage festival (top) and community members interacting with reconstructed vessels during the same event (bottom). © Rebecca Bria.

teachers working across the province of Huaylas to expand the area's history curriculum so that it includes knowledge derived from our archaeological research. In this effort, we have co-hosted workshop events for these teachers in our lab space and are co-developing learning modules for the classroom. We have also spent considerable time working in the Hualcayán school leading activities ranging from workshops on ancient iconography and 3D modelling to English lessons.

We have long sought to emplace a fully community-based participatory research (CBPR) project that centres the community's interests in the work we carry out, but local politics, and surely our own shortcomings in navigating them, have hindered our ability to proceed with consensus on a fully CBPR project at the community level. The parcelling of otherwise communal land among members has in part complicated this direction, for the community is split between those who see archaeological areas as the purview of all members that should be managed and protected communally and those who prioritize the rights of families to determine the future of any work on their parcel of land. Each season of fieldwork brings fresh challenges as we navigate these tensions, but we do not feel it is ethical to push for or force consensus. Instead, we spend time engaging in open discussion during community meetings to gather feedback on our research plans and discuss how we might work together on an initiative in the coming year. In these instances, we invite the community to take the lead on our next season of work but without forcing them to direct the aims of the project if it is not a current priority or could cause stress in the community.

We also do our best to bring humility and gratitude to our field and laboratory work, whereby our archaeological practice of care extends to our relationship with archaeological remains themselves and the ancient people whose lives we seek to know. As a team that involves many foreigners as well as non-local Peruvians, we recognize our privilege as we excavate or handle artefacts and human remains, bearing witness (cf. Hauser et al., 2018) to the lives of the Andean people who built ancient Hualcayán. We have ongoing discussions with team members and students about this privilege we have studying the lives of ancient Andean people and all that is at stake with our work, from the (mis)representation of Andean people and their history to the community's very identity and well-being. Respect is required when handling all remains and Indiana Jones-esque narratives and jokes about the ancient people we study are prohibited among students and staff alike.

Although the project has invested in many digital technologies, from drone-based 3D mapping to field databases, we also do our best to practise 'slow archaeology' (Caraher, 2019; Caraher, 2016), taking the time to collect and document artefacts with care while reflecting on our process as we progress. Mindfully running a field school and working with community members helps us achieve this in part, by prioritizing discussion over progress. Slowing down includes giving time to community members who stop by the excavations or laboratory, giving interested children the opportunity to do excavation or laboratory work with us where it is safe for them to do (Figure 14.5). Participating adults are paid for their time and collaboration. Slowing down also includes protocols such as collecting artefacts

Figure 14.5 Kintani shows up to the PIARA excavations daily after school to be a part of the archaeological team. He shows incredible skill, care, and enthusiasm in uncovering and interpreting artefacts. © Rebecca Bria.

by 1 × 1 m subunits to ensure we are gleaning the most information possible from our excavations, an inherently destructive technique.

Finally, we attempt to always keep the ancient people whose histories are uncovered through our work at the forefront of our minds and the focus of our respect. In this, we combine perspectives from community members when offered with broader knowledge about Andean ontologies to seriously contemplate the desires and wishes of both human and non-human beings, from ancestors to mountain deities, who inhabit the local landscape. This often means carrying out essential ritual acts like *pagos*, or payments to the earth, during our fieldwork, although people in Hualcayán are increasingly distancing themselves from these ideas and practices.

Because our research is undertaken with and is about people whose heritage we do not share – as a foreigner (Bria) and a Peruvian from the city of Lima

(Casanova) – we often foreground the humanistic dimensions of our findings to ensure the value of ancient lives is not lost in the data we collect and publish. One example is that we have devoted considerable time to reconstructing ancient artefacts, going through hundreds of bags and thousands of ceramic sherds over multiple seasons to find fragments that fit together and reconstruct or partially reconstruct vessels. The principal goal is to experience these vessels closer to how they would have been seen and encountered in the past. To be sure, there is an analytical dimension to this work, as reconstructions assist in our interpretation of the remains. However, we are motivated by how reconstructed vessels have greatly impacted the community's sense of interest and pride in the local archaeology as complete or semi-complete vessels are compelling for thinking through how and why people made or used them (Figure 14.4), and they make local finds seem more connected to objects displayed with honour in the provincial museum.

Given the numerous ceramic effigies of humans and animals with depictions of ancient elites, ancestors, and mummified faces in the Hualcayán collection, the reconstructed artefacts also allow us and the community to contemplate what ancient beliefs and community life were like. There are laws prohibiting the permanent display of excavated remains without a permanent representative from the Ministry of Culture, so we keep these objects curated in our laboratory and pull them out for display during open house events and when any community members stop by for a tour of the laboratory. Working with the Ministry of Culture to keep our artefacts in our Hualcayán lab house has been essential to our ability to do these kinds of activities and make objects accessible to the community.

14.4 Finding Common Ground

PIARA leadership is constantly balancing the project's role as a kind of *stakeholder*, with an interest in seeing the local past preserved, understood, and valued, and as a *partner*, with a priority to support the community as it makes decisions in the present and for the future. We hope to raise the visibility and relevance of the Andean past through our work, which has been devalued in places like Hualcayán through centuries of colonialism and nation building priorities. And yet, we have made it a policy to resist elevating the value of archaeological remains and archaeological knowledge above the needs, desires, and immediate concerns of local people, which include producing a healthy crop, ensuring the education and advancement of their children, and having enough money to put food on the table. We seek to generate a relationship of care by prioritizing the community before site preservation or research. In this, we actively try to avoid creating a patronizing relationship, such as if we the outsiders (whether from the capital city of Lima or a university in the United States) were to use money, influence, or threats of government complaints (*denuncios*) to ensure that the community protects the archaeological site however we dictate. Our strategy has been instead to put people first while building the knowledge and value of the past together – through ongoing participation in archaeological research and the many conversations we have had over many years.

Finding common ground for valuing the past has taken time, but we also recognize that forcing or expecting consensus on the meaning or importance of archaeological remains is both practically unachievable and ethically misguided. For example, some of us on the team had been trained in Andean ethnographies of scholars like Catherine Allen (2002) who documented the relationships between Andean people and the non-human beings and ancestors that inhabit the world. They include the *Machukuna*, or ancient ones, who continue to occupy the *chullpakuna*, which are ruins that include above-ground stone ossuaries (*chullpa*) where ancient human remains are still visible today. Our experiences working in other villages reflected a similar relationship, whereby our archaeological work near tombs required many ritual precautions administered by a local shaman.

In Hualcayán, which has no shortage of ancient remains and tombs, we found a very different kind of local relationship with its ruins. This leaned towards a recognition of the ancient ones, but with a kind of indifference to their presence. Local perspectives about the past vary, but in conversations with some second-generation evangelical Christians in the community we learned that such indifference may in part be driven by the perception that ancient people were pagan non-believers (often referred to as *gentiles*) and even a different race than the living community. Most individuals over 40 stopped participating in Andean traditions like ritual payments to the mother earth (*Pachamama*) in their own lifetimes during childhood. That was either when their parents intentionally turned away from these practices as part of an evangelical conversion or when they began school, which meant they began speaking more Spanish and adopted a more mestizo identity.

This generation had come of age about 20 years before PIARA's arrival in Hualcayán, and there wasn't an overwhelming curiosity to reclaim Indigenous knowledge or even to be aware of ancient remains, except on occasion. While we arrived to Hualcayán ready to be partners in valuing the local past, we had to accept that local people's interest in the archaeology did not meet or exceed our own – even after our early work revealed that the site was one of highland Peru's earliest settlements with monumental temple constructions dating back to 2400 BCE.

While respecting the people of Hualcayán's autonomy and self-determination, our conversations and outreach initiatives have attempted to provide opportunities for people to explore the relevance and role of archaeology in their community on their own terms while working to honour the lives of the ancient people we study. As scholars of heritage, we recognize archaeological remains are finite resources that might support several untapped pathways for the community as it faces the future, whether by increasing tourism or enriching a sense of place, history, and identity. Also behind these initiatives is a recognition that the devaluing of Andean histories has been centuries in the making: unjustly diminished and utterly transformed by colonialism and its historical trauma. Nonetheless, we are careful not to dictate how local people should value the past.

The PIARA team sees itself as on a journey to discover that value alongside the people of Hualcayán as knowledge is produced, whether at the trowel's edge, in the laboratory, during private conversations, or at community meetings and public events. This collaborative journey extends to children in the local school (Figure 14. 5) where we have carried out numerous activities, including using storytelling and artwork to bring their own vision of the past to life using the knowledge they gain participating with us in the archaeological steps of observation, digital documentation, interpretation, and engagement with objects and spaces (Bria and Casanova Vásquez, 2022).

14.5 Collaborative Initiatives to Revalue the Past, Build Tourism, and Address the Climate Crisis

We are experimenting with several collaborative initatives to enhance our discussions about the past. One initiative is to create digital reconstructions of local ancient structures and artefacts. Most spaces and objects are uncovered only partially through excavation, and 3D reconstructions make it possible to experience them more completely, especially when using virtual reality (VR). With a VR headset and controllers, users can be digitally immersed in another time or place. The prototype VR experience we recently created, which was led by our expert collaborator Bruce Carlisle, features a digital reconstruction of one of Hualcayán's most monumental ancient spaces: a Preceramic Period temple that was part of a religious tradition called Kotosh (2400–1800 BC; Figure 14.2, bottom). In this experience, users participate in a virtualized practice of picking up a cob of corn – a ritual food in the early Andes – before ascending to the top of the temple and depositing the corn in a sacred fire at the centre of the temple. This experience featured the sound of rushing water captured from the village waterfall that can be seen in the distance and the crackle and spark of fire when the offering was made. When community members testing the VR experience looked around from inside the digitally reconstructed temple, they instantly recognized where they were within the village landscape, allowing users to connect the past to the present (Figure 14.6).

Our discussions about the ancient importance of this temple for the early prehistoric community at Hualcayán were instantaneously richer than we had previously experienced, even after having community members tour our excavations. It gave us a common, albeit simulated, experience to think through what it might have felt like to be an ancient person in Hualcayán. We are sensitive as to whether and how such digital tools, which originate from Western ways of seeing and rendering the world, may manipulate rural Andean perceptions in a negative way. But we approach the VR experience as a way to make archaeological knowledge both experiential and rooted in storytelling, which are essential elements in Andean ways of knowing, albeit in a different form.

Our time spent creating the VR experience has provided a way to connect with community members through archaeology and generate a collaborative activity whereby the community takes a lead in helping to reimagine the past. We are continuing to modify the VR prototype with community feedback, and with recent funding from the US National Endowment for the Humanities, we are now collaborating

Figure 14.6 Community elders and young adults attend an event in the PIARA laboratory where they use VR headsets to explore a local ancient temple that PIARA reconstructed in 3D. © Rebecca Bria.

to create a full VR experience and video game about ancient Hualcayán that will demystify the complex seven cultural phases of the site's pre-colonial past. We are mutually excited about this endeavour, which will provide a focal point for us to bring together the Hualcayán community's vision of how Hualcayán's predecessors lived with our own. If successful, it would bring Hualcayán to the attention of people around the world who play the Hualcayán game, as well as support the community financially through game sales or even site visits.

Two additional collaborative projects, which are still in their initial planning stages, have less to do with advancing archaeological research and more to do with leveraging our knowledge, skills, and networks as archaeologists to support community needs. These are: (1) to design and build a town plaza that reflects the community's history and identity with interactive information about the local archaeology for visitors and (2) to mitigate the threats of climate change and glacial retreat on local farmers.

The plaza project has turned out to be one of our most unexpected but exciting collaborations. In Peru, plazas have long been essential components of the built environment, making and marking place and providing a venue for asserting political identity (see Hirsch, 2016, inter alia). Hualcayán has never had a plaza – only an open square field at the centre of town – and community members express some embarrassment about the state of it. Roads in Hualcayán are primarily graded earth, but the community has recently invested in a concrete pavement and sidewalk around the central square. They are now looking to create a proper plaza that reflects who they are. They have asked us to partner with them to help make it a reality and are interested in incorporating symbols and iconography from the archaeological site into the design. Working on a budget, we reached out to an ex-project member in the United States who is now working professionally as an architect. Although his time working in Hualcayán ended ten years ago, he jumped at the chance to design the plaza with the community, and we have been working through ideas, sketches, and maps to bring it into existence in the coming year. While the VR project gave us a medium and focal point for exploring what mattered to the ancient Hualcayán community, the plaza project is an opportunity for us to support the community as they decide what about that past is meaningful to their identity today.

We are also approaching the plaza project as part of a broader plan to establish greater infrastructure for tourism in the community. Hualcayán lies in the buffer zone of the Huascarán National Park – a World Heritage site for its unique tropical glacier ecosystem – and has a dedicated base camp for visitors embarking on one of the park's 37 treks. Limiting the number of visitors in Hualcayán, which total around 60–100 of the total ~200,000 who visit the park each year (15% foreigners; Ministerio de Comercio Exterior y Turismo del Peru, 2024), is the fact that the Hualcayán base camp is the farthest from the regional capital of Huaraz where most visitors link up with trekking companies. Moreover, the trek from Hualcayán is more challenging than popular alternatives in the park. Nonetheless, the community's interest in developing archaeological tourism at Hualcayán seeks to leverage their resources to give people an additional reason to visit, to stay longer, or to choose Hualcayán's trek over others.

The PIARA team is working with the community to develop ways to bring more robust information about Hualcayán's history to visitors, and we will use the plaza as a starting point for learning about and exploring the archaeological site. Specifically, we will use traditional signage that is creatively mixed with object-based augmented reality (AR), whereby 3D experiences and information will be triggered by pointing smartphones at archaeology-themed concrete sculptures planted throughout the plaza and potentially a GPS-guided tour of Hualcayán's archaeology. While the early stages of the plaza project are funded by portions of PIARA's annual research budget (a budget that currently originates from a mix of small grants and field school fees), we are in the process of submitting major grants to develop the AR and guided tour activities.

The second project underway challenges the team to leverage our knowledge about ancient agricultural practices and infrastructure and apply it to the many pressing issues farmers face due to climate change. Communities like Hualcayán face many risks linked to the rapid melt of the tropical glaciers that provide water across Peru, such as shifting watersheds, landslide risks, water contamination, and overall stress on agricultural yields (Lozano-Povis et al., 2021; Bury et al., 2013; Mark et al., 2010; Motschmann et al., 2020).

While adapting to a changing climate, the people of Hualcayán are also managing environments heavily shaped by the infrastructures of their ancestors. While such anthropogenic landscapes both constrain and enable farming success in various Andean regions (Erickson, 2003; Erickson, 1998; Kendall, 1994; see also Robbins, 2019), people's use and management of this ancient infrastructure are only beginning to be studied in highland Ancash, including by contributors to this volume (Herrera, 2019; Lane, 2017; Bria and Walter, 2019; Branch et al., 2023; Conlon and Lane, 2019). We are looking to build on this work by creating a study that employs a CBPR methodology to bring together local/Indigenous knowledge with scientific research on climate change, infrastructure, and decision-making to identify ways to mitigate the inevitable effects of climate change in the region.

While funding is pending on this project, advancing this collaborative work has become a priority for the PIARA team because of its urgency to the current well-being and future livelihoods of our community partners in Hualcayán and surrounding villages. In collaboration with ethnographers Doris Walter, Jessica Eise, and Arnaldo Castillo Garcia, we have conducted a series of interviews over several years to carefully understand people's experiences of a changing climate and their relationship with ancient infrastructure as we embark on this work. This research is part of a broader effort by anthropologists globally to not only confront the impacts of the climate crisis on archaeological heritage but also consider how we can meaningfully leverage archaeological knowledge to address these urgent issues (see d'Alpoim Guedes et al., 2021, p. 905–906).

14.6 Reflections: Learning from Our Challenges, Successes, and Failures

As she navigated academia as an early career scholar, the first author has often been discouraged from investing much time and effort on community collaborations, even while being congratulated for it. During her dissertation, she often heard, "This is great, but why don't you wait until you get your PhD to put together a community-engaged project? You'll never finish your degree if you keep doing all this community work." Or, once the PhD was acquired, "Why don't you wait until you get tenure to worry about community initiatives? If you don't get tenure, you won't be able to support the community because you'll be out of a job."

From the perspective of academia, these comments are completely right. If she failed/s to succeed in academia, the dream of impactful archaeological work in a community she cares about would likely diminish, if not go away entirely. Although it has been nearly a decade since McAnany and Rowe (2015) declared a paradigm shift towards collaborative archaeology was underway, there are still challenges, especially to early career scholars and projects, to supporting the slow, careful work that is needed to produce meaningful, heart-centred research. Archaeologists engaged in collaborative work must often push forward their convictions despite real pressures to find paths of least resistance; this, too, is an act of care.

It is both moving and agonizing to consider our impact in Hualcayán over the past 15 years. We have spent a great deal of this time worrying about whether we are going about things the wrong way or feeling as if there are insurmountable problems as we confront community troubles or threats to the archaeological site. For example, only a year into our work at Hualcayán, when we were just beginning to navigate how to be stewards of Hualcayán's archaeology while supporting the community's interests, a Peruvian company arrived to build a hydroelectric canal. This project would both bring construction jobs for five years and destroy hundreds of ancient tombs, while also diverting a large portion of the community's water away from the village. After a period of intense and prolonged negotiation, the community eventually denied the project. Nonetheless, such challenging moments have given us consistent opportunities to define our relationship with the community and their archaeological remains. We have always strived to put the community's needs first, even if it means some archaeological contexts become compromised, while also advocating for the mostly unknown past hidden within Hualcayán's soils – a balancing act that must proceed with care at every turn.

Having maintained a presence in Hualcayán for over a decade, both sides of the partnership have weathered many storms and seem to have moved from mutual feelings of uncertainty to a comfortable place of trust. And yet, having worked on other projects before embarking on this long-term project at Hualcayán, we have learned that the longer you work in a community, the higher the stakes become. A two-year project with clear goals and a defined community payout or end product may begin and end smoothly, as no one expects much of the other in this positive but transactional relationship, even if the arrangement is mutually beneficial: the archaeologist acquires local workers to advance their research, and the community can rely on paid labour for a few seasons. In contrast, we have learned that with a long-term project both sides start to expect more from each other, as they should. The relationships and thus obligations change as our lives become more intertwined.

The height of the COVID-19 pandemic between 2020 and 2022 was a difficult time, but it offered the PIARA leadership and the community of Hualcayán some healthy distance to consider our relationship with one another. Did the community still want us there after such a long break away? Did our absence have any impact, positive or negative? Did anyone in Hualcayán even think about archaeology once we left? The pandemic commenced around a decade after our first arrival in Hualcayán, and these questions overshadowed our return in 2022. Yet it seemed that our time away allowed us to reflect on and then renew and strengthen our partnership and mutual commitment, although this may be more a testament to surpassing the ten-year mark in our relationship and deepening the ties we have. Regardless, now more than ever before the community is coming to us to initiate conversations about the future of their archaeological remains in terms of preservation, tourism, and other hopes.

Many children we spent time with at the community school a decade ago are now young adults, some having been elected to the community leadership council (*junta directiva*). Community elders are excited by the new initiatives we are working towards around water and farming and are more open to discussing archaeology in their community. Many community members have expressed interest in finding better ways to ensure all community members are preserving the archaeological remains on their allotted lands so that we can actively work towards a future with archaeological tourism in Hualcayán, although community members are generally cautious not to interfere with the farming decisions of other families. We are using our connections at UNASAM to establish a *convenio* directly between the university and the community, making it easier for young adults from Hualcayán to be admitted to college and providing a way for the community to reach out directly to experts in health, education, and agriculture without the need for our intervention or mediation.

Many challenges nonetheless likely await our relationship. Glacial retreat is putting pressure on the water supply and the community recently built a road to carry heavy tubing and building supplies up the mountainside, causing landslides to cover or destroy important ancient tombs. The move to install drip irrigation systems to conserve water throughout Hualcayán's fields has been extremely positive for crop growth, but has led to the construction of many reservoirs, some of which have destroyed ancient buildings. Growing families in need of land are expanding into archaeological features.

The new interest in archaeological preservation at the community level has also led to troubles and tensions for certain community members whose agricultural land is on or near the more significant areas of the archaeological site, leading to new tensions with our oldest and closest collaborators. They are not at fault for using this land, for it was parcelled out 30 years ago, long before there was talk of archaeology and its importance. Other families don't want to give up their land to make room for others to stop ploughing these significant areas, yet they want the archaeological destruction to stop. These situations are straining our personal relationships, because many have become our close friends or *copadres* (co-parents through a godparent relationship). These farmers may be told by neighbours that they cannot have a say in what happens to their fields because it is 'the archaeologists' land,' whereby the PIARA team is held up as disenfranchising the farmers, even if we have not actively pushed to stop their farming practices. Only time will reveal whether the trust we have built and/or the previous and current acts of care will bring us to a resolution. Perhaps our persistent challenges present a principal lesson: just like any relationship, a community partnership will be in a constant state of change as needs and desires change, and one must navigate them from the heart.

Note

1 Project leadership includes Rebecca Bria and co-directors Cora Rivas Otaíza (2009–2010), Felipe Livora Castillo (2010–2013), Elizabeth Cruzado Carranza (2013–2015), and Erick Casanova Vasquez (2018–present). Many others have contributed significantly to the project.

References

Allen, C. J. (2002). *The Hold Life Has: Coca and Cultural Identity in an Andean Community.* Washington D.C.: Smithsonian Institution Press.

Branch, N., Ferreira, F., Lane, K., Wade, A., Walsh, D., Handley, J., Herrera, A., Rodda, H., Simmonds, M. and Meddens, F. (2023). Adaptive Capacity of Farming Communities to Climate Change in the Peruvian Andes: Past, Present and Future (Preliminary Findings of the ACCESS Project), *Revista de Glaciares y Ecosistemas de Montaña,* 8, pp. 51–67.

Bria, R. E. (2017). *Ritual, Economy, and the Production of Community at Hualcayán, Peru (Ph.D. Dissertation).* Vanderbilt University.

Bria, R. E. and Casanova Vásquez, E. (2022). 'Digital Archaeology and Storytelling as a Toolkit for Community-Engaged Archaeology', in: K. Garstki (ed.) *Critical Archaeology in the Digital Age.* Los Angeles: UCLA Cotsen Institute of Archaeology Press.

Bria, R. E. and Cruzado Carranza, E. K. (2015). Making the Past Relevant: Co-Creative Approaches to Heritage Preservation and Community Development at Hualcayán, Ancash, Peru. *Advances in Archaeological Practice,* 3, pp. 208–222.

Bria, R. E. and Walter, D. (2019). Multiple Ways of Understanding Peru's Changing Climate, *Open Rivers: Rethinking Water, Place & Community*, 14, pp. 11–25.

Bury, J. T., Mark, B. G., Carey, M., Young, K. R., Mckenzie, J. M., Baraer, M., French, A. and Polk, M. H. (2013). New Geographies of Water and Climate Change in Peru: Coupled natural and social transformations in the Santa River watershed, *Annals of the Association of American Geographers,* 103, pp. 363–374.

Caraher, W. (2016). 'Slow Archaeology: Technology, Efficiency, and Archaeological Work', *in:* E. W. Averett, D. Counts and J. Gordon (eds.) *Mobilizing the Past for a Digital Future: The Potential of Digital Archaeology.* Grand Forks, ND: The Digital Press @ The University of North Dakota.

Caraher, W. R. (2019). Slow Archaeology, Punk Archaeology, and the 'Archaeology of Care', *European Journal of Archaeology,* 22, pp. 372–385.

Ccahuana Córdova, J. A. (2020). La reforma educativa de 1905: Estado, indígenas y políticas racializadas en la República Aristocrática, *Apuntes,* 47, pp. 5–32.

Conlon, S. and Lane, K. J. (2019). Peru's Ancient Water Systems Can Help Protect Communities from Shortages Caused by Climate Change. *The Conversation*. Available at: https://theconversation.com/perus-ancient-water-systems-can-help-protect-communities-from-shortages-caused-by-climate-change-121284

Connolly, R. P., Bria, R. E. and Cruzado Carranza, E. K. (2015). 'Co-Creation and Sustainable Community Engagement', in: J. Decker (ed.) *Innovative Approaches for Museums: Collections Care & Stewardship.* Rowan & Littlefield.

D'alpoim Guedes, J., Gonzalez, S. and Rivera-Collazo, I. (2021). Resistance and Care in the Time of COVID-19: Archaeology in 2020, *American Anthropologist,* 123, pp. 898–915.

De La Cadena, M. (2000). *Indigenous Mestizos: The Politics of Race and Culture in Cuzco, Peru, 1919-1991.* Durham: Duke University Press.

Erickson, C. L. (1998). 'Applied Archaeology and Rural Development: Archaeology's Potential Contribution to the Future', *in:* M.B. Whiteford and S. Whiteford, S. (eds.) *Crossing Currents: Continuity and change in Latin America.* Saddle River, NJ: Prentice Hall.

Erickson, C. L. (2003). Agricultural Landscapes as World Heritage: Raised Field Agriculture in Bolivia and Peru. *Managing Change: Sustainable Approaches to the Conservation of the Built Environment*, pp. 181–204.

Hauser, M. W., Battle-Baptiste, W., Lau-Ozawa, K., Voss, B. L., Bernbeck, R., Pollock, S., Mcguire, R. H., Rizvi, U. Z., Hernandez, C. and Atalay, S. (2018). Archaeology as Bearing Witness (Vital Topics Forum). *American Anthropologist,* 120, pp. 535–548.

Herrera, A. (2019). 'Indigenous Technologies, Archaeology, and Rural Development in the Andes: Three Decades of Trials in Bolivia, Ecuador, and Peru', i*n:* C. Isendahl and D. Stump (eds.) *Oxford Handbook of Historical Ecology and Applied Archaeology.* Oxford: Oxford University Press.

Hirsch, E. (2016). Mediating Indigeneity: Public Space and the Making of Political Identity in Andean Peru, *PoLAR: Political and Legal Anthropology Review,* 39, pp. 95–109.

Kendall, A. (1994). *Proyecto arqueológico Cusichaca, Cusco: investigaciones arqueológicas y de rehabilitación agrícola, Lima, Southern Peru copper corporation.* Southern Peru Copper Corporation.

Lane, K. (2017). Water, Silt and Dams: Prehispanic Geological Storage in the Cordillera Negra, North-Central Andes, *Peru, Revista de Glaciares y Ecosistemas de Montaña,* 2, pp. 41–50.

Lozano-Povis, A., Alvarez-Montalván, C. E. and Moggiano, N. (2021). Climate Change in the Andes and Its Impact on Agriculture: A Systematic Review, *Scientia Agropecuaria,* 12, pp.101–108.

Mark, B. G., Bury, J., Mckenzie, J. M., French, A. and Baraer, M. (2010). Climate Change and Tropical Andean Glacier Recession: Evaluating Hydrologic Changes and Livelihood Vulnerability in the Cordillera Blanca, Peru, *Annals of the Association of American Geographers,* 100, pp. 794–805.

Mcanany, P. A. and Rowe, S. M. (2015). Re-Visiting the Field: Collaborative Archaeology as Paradigm Shift, *Journal of Field Archaeology,* 40, pp. 499–507.

Ministerio de Comercio Exterior y Turismo del Peru. (2024). *Áncash Reporte Regional de Turismo*, Año 2023.

Motschmann, A., Huggel, C., Carey, M., Moulton, H., Walker-Crawford, N. and Muñoz, R. (2020). Losses and Damages Connected to Glacier Retreat in the Cordillera Blanca, Peru, *Climatic Change,* 162, pp. 837–858.

Robbins, P. (2019). *Political Ecology: A Critical Introduction*. John Wiley & Sons.

Supernant, K., Baxter, J. E., Lyons, N. and Atalay, S. (2020). *Archaeologies of the Heart.* Springer.

15 Utilising Cultural Heritage to Improve Water Security and Agro-Pastoral Farming in the Peruvian Andes

Kevin Lane, Nicholas Branch, Josephine Handley, Frank Meddens, Pedro Gonzalez, Douglas Walsh, Mario Advíncula, Alexander Herrera, Cirilo Vivanco and Wilmer Moncada

15.1 Introduction

Across the Peruvian Andes, the significance and value of high-altitude wetlands (lakes, peatlands, and wet meadows over 2500 m a.s.l.) and their associated water management infrastructure (e.g. micro-dams) should not be underestimated. For millennia, these have provided an essential contribution to a range of ecosystem services, especially as a source of water for cultivation and highly nutritious grazing land for animals, in addition to sequestering carbon, regulating water flow and enhancing biodiversity. Farming communities, and non-governmental organisations (NGOs) working with communities, recognise their importance and play an essential role in their management and conservation. Threats to wetlands due to climate change, including loss of glacier meltwater recharge, as well as drainage, conflict and mining, are well recognised. As such, these threats pose a considerable concern for water security and agro-pastoral farming.

In this study, we summarise past and present water management in the Peruvian Andes and then undertake two case studies of wetlands and their associated water management infrastructure located in central Peru (Figure 15.1). The first is Antaycocha (Chillón valley, Lima Region), where we demonstrate how palaeoecology can be used to inform about the timing and role of dam construction and associated socio-economic change during the last 3000 years. Our second study concerns Ricococha Alta (Cordillera Negra, Ancash Region) where we demonstrate the contribution of indigenous knowledge and community engagement in dam restoration for protecting cultural heritage and sustainable development in the context of climate change. These studies show how a blend of knowledge from cultural heritage, palaeoecology and indigenous communities can be used to better understand human–wetland interactions and provide know-how to inform approaches to sustainable rural development.

DOI: 10.4324/9781003491071-18

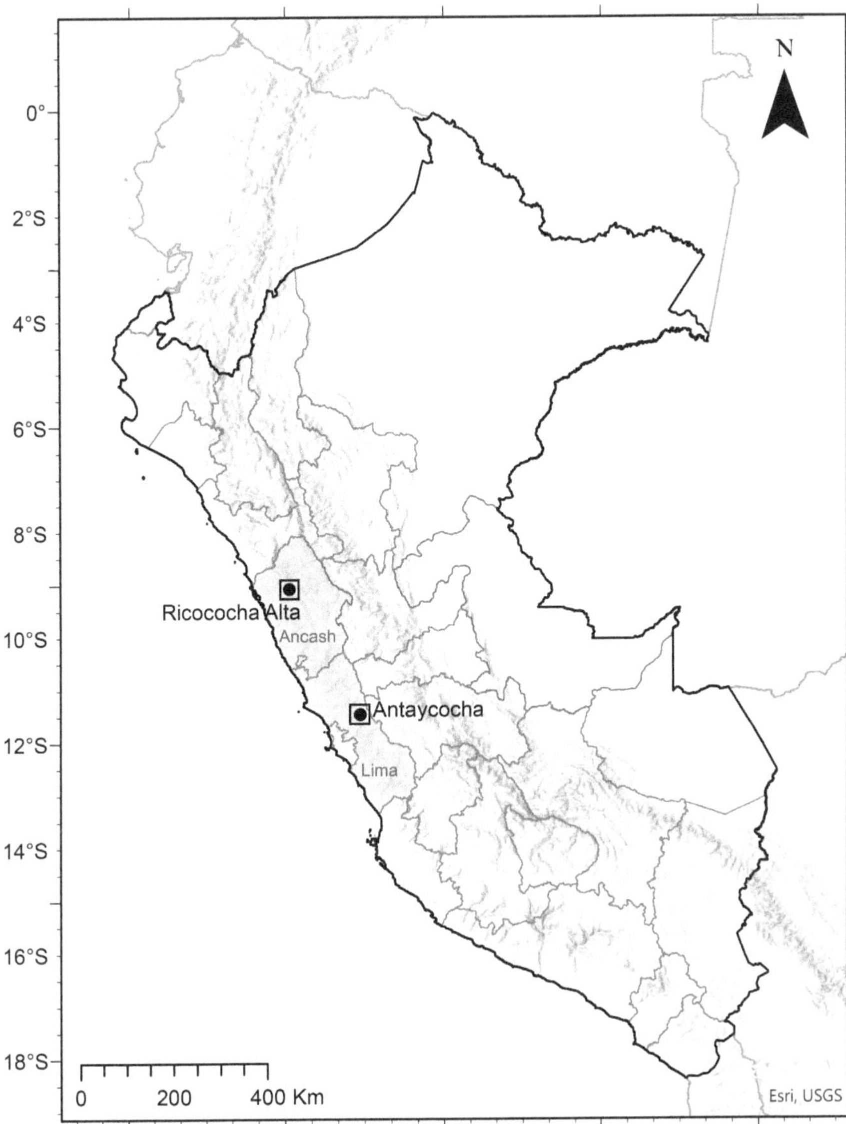

Figure 15.1 Location of Ricococha Alta (Ancash) and Antaycocha (Lima), Peru. © Josie Handley, www.gadm.org/

Note: Produced in ArcGIS Pro 3.2.0, using GADM (2018) administrative boundary data.

15.2 Water Management in the Peruvian Andes, Past and Present

In the 21st century, water scarcity is increasingly seen as the main threat to human world economies and the harbinger of future conflict (Gleick, 1993; Swain, 2015; Patel et al., 2022). In this regard, the Peruvian Andean highlands constitute one such crisis region where lack of water is understood by experts as the single most threatened natural resource in the face of climate change and ever-retreating tropical glaciers, a sentiment echoed by local governments, NGOs and especially rural communities (Stark et al., 2012). Yet the present-day water scarcity is not a new phenomenon.

Archaeological, palaeoenvironmental and palaeoclimatic evidence suggests that climate change and water insufficiency were recurrent issues throughout Andean prehistory. One such period of climate instability is connected to the late Middle Horizon through to the Late Horizon when the Medieval Climate Anomaly (MCA: AD ~900–1300) and subsequently the Little Ice Age (LIA: AD ~1325–1880) impacted the region (e.g. Haug et al., 2001; Rein et al., 2005; Apaéstegui et al., 2014; Stansell et al., 2017; Lüning et al., 2019). During this time, especially the MCA, lower precipitation and higher temperatures in warmer seasons undoubtedly meant less water.

Against this backdrop, pre-Hispanic communities across the Andean highlands invested heavily in hydraulic technology, thereby guaranteeing local resilience (e.g. Grana et al., 2024; Percero-Oubiña et al., 2016; Lane, 2021; Vining and Williams, 2020; Salminci et al., 2014). Part of this Pacific-facing mountain range is the Cordillera Negra in the Ancash highlands, North-Central Peru. Here, community-led investment in drainage-wide, integrated agro-pastoralist hydraulic infrastructure, including dams, check dams, silt traps, terraces, reservoirs and canals increased substantially during the Middle Horizon (AD 600–1000) through to the early Spanish colonial period (AD 1536–1615; Lane 2009). Indeed, such was the investment in hydraulic technology and the concomitant increase in human population among the local Huaylas population (Espinoza Soriano, 2013) that we can perhaps understand their adaptation as a study in ecological anti-fragility (Equihua et al., 2020).

In the present day, water access and security is a perennial worldwide problem critically impacting Peru (Ministerio de Desarrollo Agrario y Riego, 2019). A local adaptation to the increasing water shortage in the Andean region is the construction of micro-dams using concrete and steel. As of 2015, 743 micro-dams had been built in the Andean area, substantially adding to water security in highland rural areas. Micro-dams control water and soil erosion, feed subterranean aquifers and provide for people's livelihoods. Approximately 75% of these are at high altitude and directly impact the water supplies of local rural populations and then feed downstream to coastal regions (Autoridad Nacional del Agua, 2015).

Modern micro-dam construction is seen to tackle climate change-induced water scarcity, and through it, rural land abandonment, crop failure, outward migration, and community cohesion breakdown. Most of these dams have been built in the past 40 years (Paredes-Beltran et al., 2021). At a local highland environment level, the construction of micro-dams and reservoirs is seen as an important mitigating factor against climate change and water insecurity (Llosa Larrabure, 2008), especially given accelerating glacier retreat (Proyecto Glaciares+, 2018).

In many cases, however, the use of modern materials is not a sustainable or optimal solution. Cement and steel dams are prone to damage during seismic events, and once damaged are impossible for local communities to repair, leading to a functional lifespan of only 50 years (Wieland, 2010). They are very expensive and time-consuming to build, requiring non-local materials and constructors secured through local authorities in a process open to corruption. Furthermore, modern dam construction in the Andes mimics ancient adaptations to the same problem. Therefore, many of these new dams are built over ancient structures irrevocably destroying them; most of the pre-Hispanic structures are over 500 years old and can help us understand how ancient populations adapted to past climate change. Over 95% of these structures are currently unregistered archaeologically. Given the increasing water stress in the Andes, the current rate of ancient dam loss and destruction is unprecedented. Moreover, these ancient dams constitute available *installed capacity,* which can be rehabilitated for modern use.

15.3 Palaeoecological Investigation of the Antaycocha Wetland, Chillón Valley, Lima

15.3.1 Site Location and Archaeology

Antaycocha wetland (18L W: 8734617.13/S: 32700.00 UTM (WGS84): 3601 m a.s.l.) is situated in the Chillón Valley near the village of Canta. The wetland has a pre-Hispanic stone micro-dam feature at one end (Figure 15.2) and is fed by the Rantao Stream at the other, a water course originating from a spring in the mountain above (Cerro Paraguay). Modifications have been made to the stream's channel above the wetland in the form of a series of stone steps to slow down the flow of water in times of heavy rain (Farfán, 2011).

The Cantamarca archaeological site is situated on the hill above Antaycocha and is one of the main Late Horizon Inca sites in the valley. Although its origins are older and are thought to date to the first Lima occupations in the Early Intermediate Period (AD 1–650), evidence to support continuous occupation prior to the Late Intermediate Period at the site is scarce (Farfán, 2011). Populations began to expand in the Late Intermediate Period with the development of high-elevation defensive settlements on ridgetops. During this time, there was limited social stratification; instead, multiple political units controlled small territories and sources of irrigation water, with evidence for sites expanding and developing over time with no formal planning. During the Late Horizon, there was evidence of significant state influence from the Inca, who exerted control over large portions of the valley probably to control water resources. Sites at higher elevations were closer to main water sources (e.g. lakes and glacial meltwater) and provided the Inca not only with control of water but also with cultivation in the lower valley's desert-like areas.

Figure 15.2. The pre-Hispanic stone micro-dam on the Rantao stream at Antaycocha. © Nicholas Branch.

15.3.2 The Palaeoecological Record

To understand the developmental history of the Antaycocha wetland, a 5.97 m deep sediment core was recovered from the basin (Figure 15.3). The radiocarbon dating of the sequence revealed the sediments had a basal age of ~1010 BC (2960 cal. BP), with sediments accumulating over nearly the last 3000 years. Surprisingly, most of the sequence dates to only the past 550 years, with sediments located at 5.5 m dating to AD 1410–1455 (540–495 cal. BP) (a late, Late Intermediate Period to Late Horizon date). Consequently, a large volume of sediment has accumulated in the wetland over a relatively short time.

Within the basal 50 cm of the sequence, which pertains to a period of 2500 years largely covering the Early Horizon (800–200 BC; 2750–1850 cal. BP) and the Early Intermediate Period (200 BC–AD 650; 1850–1300 cal. BP), the record consists of inorganic lake sediments. The paucity of sediment accumulation during this time was likely due to stream flow into the basin having a higher rate. Within the pollen record (Figure 15.3), there are some indicators of human activity during the Early Horizon and Early Intermediate Period (LPAZ1), including the presence of *Sporormiella*, a fungus that grows on herbivore dung. Its occurrence within the

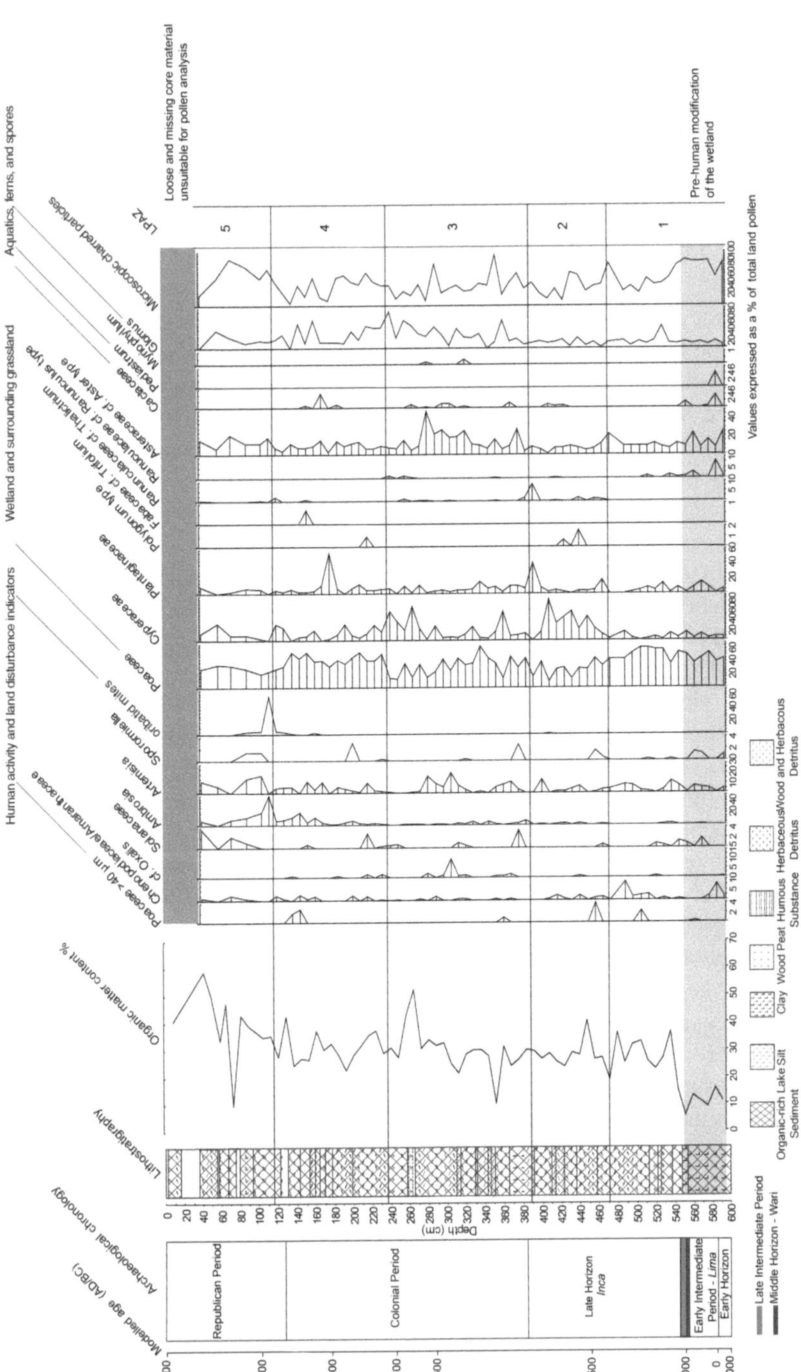

Figure 15.3 Antaycocha palaeoecological data together with the age-depth model and archaeological chronology.© Josie Handley.

Note: Diagram produced using TILIA and TILIA.GRAPH (Grimm, 1991).

Early Horizon and Early Intermediate Period probably reflects seasonal herbivore (camelid) use of the wetland for grazing and drinking water (see Chepstow-Lusty et al., 2019). Together with pollen evidence for Chenopodiaceae/Amaranthaceae and Solanaceae, it may indicate early agro-pastoral farming practices taking place around the wetland with both plant families containing important Andean crops, i.e. quinoa (*Chenopodium quinoa)* and potatoes *(Solanum tuberosum)*, respectively.

Following the Early Intermediate Period, there is a very short section of the record that correlates with the Middle Horizon (552–553 cm: AD ~751–979; ~1199–971 cal. BP). The reduction in sediment accumulation may have resulted from a lack of human activity and a consequent decline in landscape erosion within the basin catchment. This theory is supported by the absence of Wari-style ceramics at Cantamarca (Farfán, 2011), as well as a decrease in *Sporormiella* (Figure 15.3). Another possible explanation for the low sediment accumulation during the Middle Horizon could be a reduction in precipitation during the MCA (AD 900–1300; ~1050–650 cal. BP) causing less surface run-off from the surrounding slopes.

There is a renewed deposition of sediments from ~AD 1434 (~516 cal. BP), alongside an increase in organic lake sediments, which suggests a fall in the rate of water flow (Figure 15.3). The late, Late Intermediate Period/early Inca date coincides with the likely permanent occupation of Cantamarca. The marked change in lithology likely represents a period of modification of the basin, involving the construction of a stone micro-dam. The construction created a reservoir probably to increase water storage capacity mainly for agricultural practices. The timing of the reservoir construction coincided with a period of deteriorating climate during the Little Ice Age. This generally resulted in higher precipitation across the Peruvian Andes during the late, Late Intermediate Period and Late Horizon (Kanner et al., 2013; Apaéstegui et al., 2014). The construction of additional reservoirs recorded in the area may represent a response by the local populations to increased precipitation allowing for optimised water conservation practices (Farfán, 2011). Certainly, the construction of the reservoir at Antaycocha would have significantly increased the availability of water at a time when populations were growing in the valley.

Alongside the evidence for the construction of a micro-dam, there is also evidence for increased human activity at the transition to the Late Horizon (LPAZ1: AD ~1438; 512 cal. BP). Landscape disturbance becomes more apparent with soil erosion indicators, such as *Artemisia* and *Glomus*, alongside the presence of several crop types that may have been grown on the terraces. The presence of Poaceae >40 μm within LPAZ1 and LPAZ2 probably represents Late Horizon *Zea mays* cultivation, which may have been grown in association with quinoa, kañiwa and kiwicha, all pseudocereals in the Chenopodiaceae/Amaranthaceae family grown in the highlands both today and in pre-Hispanic types. The occurrence of *Oxalis* in LPAZ1 and LPAZ2 during the Late Horizon indicates that its tubers, known as oca, may also have formed an important dietary staple. In addition to crops, the presence of *Sporormiella* suggests that localised camelid husbandry also benefited from the increased water storage, probably using the reservoir for seasonal drinking water and grazing.

Following this, the transition to the Colonial Period (LPAZ3, AD 1532; 418 cal. BP) witnessed a reduction in the maintenance of the landscape. An increase in shrubland taxa (e.g. Asteraceae and Cactaceae) may signal their colonisation of the terraces and surrounding slopes. This is supported by a drop in Quinoa (*Chenopodium quinoa*) and Maize (*Zea mays*) cultivation during the first half of the Colonial Period (LPAZ3; Figure 15.3). Although there may have been a decline in cultivation, there is still a presence of animal husbandry, indicated by *Sporormiella* within LPAZ3 and 4. This likely represents a mixture of camelids and old-world domesticates such as cows and sheep, with the corrals around Antaycocha, most likely built in the Late Intermediate Period or Late Horizon, still in use today. With a fall in population levels, maintenance of the reservoir reduced and coupled with increased erosion led to the accumulation of a large amount of sediment over the past 500 years. The transition to the Republican Period at AD 1826 (124 cal. BP) recorded an increase in Oribatid mites, an indicator of herbivores in the landscape, alongside a later increase in *Sporormiella* at ~AD 1910 (~40 cal. BP; LPAZ5), providing clear evidence for herbivore activity around the wetland from the late 19th century onwards.

15.3.3 Interpretation

The timing of wetland formation at Antaycocha is of interest because the basal radiocarbon chronology indicates that sediment started to accumulate within small natural basins, sometime before ~1000 BC (~2955 cal. BP). However, once established, the wetland has evidence for human interference in its ecohydrological dynamics through the construction of micro-dams. This is dated to the Late Intermediate Period or early Late Horizon (Inca) date of AD ~1434 (~516 cal. BP) based on the radiocarbon chronology. The quality of engineering involved in the construction of the micro-dam suggests that the local community was knowledgeable and had expertise in the selection of suitable raw materials. Its construction at a time of an increasing usage of terrace agriculture is suggested by the archaeological record; such agriculture would have necessitated a dependable water supply for irrigation. The persistence of such agriculture in the landscape for over 500 years is testimony to the robustness of the micro-dam and the quality of its construction.

The archaeological evidence suggests that the timing of micro-dam construction coincided with possible agricultural expansion. This development, perhaps coupled with demographic change (e.g. migration), followed a prolonged period of lower precipitation during the MCA (AD ~900-1300; ~1050–650 cal. BP). The combination of a micro-dam, wetlands and ancillary technology (including canals, terraces and reservoirs) at Antaycocha and elsewhere in the central Andean Highlands strongly suggests that pre-Hispanic populations envisioned hydraulic management of whole watersheds integrating and feeding into agro-pastoralist economies (Lane, 2021).

Nowadays, the emphasis has changed. The Spanish colonial collapse of pre-Hispanic herding, which cemented itself in the 20th century (Flores Ochoa, 1980), has meant that in many areas of the Andes a whole way of life involving the intense rearing of domesticated camelids has practically disappeared. Against this backdrop, modern rural communities clamour for water from micro-dams which feed directly into a predominantly farming economy. This is the case of the Ricococha Alta restoration project, where the water thus gained will be used primarily for cash crop cultivation.

15.4 The Ricococha Alta Micro-Dam Restoration Project

The rehabilitation of a pre-Hispanic dam was the centrepiece of a pilot project: *Past Water Futures*. The dam in question is the Ricococha Alta dam located at 18L W: 8996716.00/S: 19304.00 UTM (WGS84) at an altitude of 4,560 m a.s.l. in the Pamparomas District, Huaylas Province, of the Ancash Region (Figures 15.1 and 15.4). This was likely first constructed in the late Middle Horizon (AD 750–1000), although there seems to have been considerable continued investment during the late, Late Intermediate and Inca Period (Lane, 2011).

15.4.1 Construction and Restoration

The dam is a double-walled construction oriented roughly east-west with a length of 45.7 m, an average width of 2 m and a maximum preserved height of 2.35 m. Both ends of this gravity dam are anchored onto the andesite bedrock. The inner and outer walls of the dam are made up of irregular stone blocks held together with a sandy clay mortar. The dam core – the infill between the walls – is composed of this same clay mixed with gravel. The dam presents a single sluice

Figure 15.4 The pre-restoration landscape of the micro-dam at Ricococha Alta. Past Water Futures Project.

at the base of the wall (17.4 m from the western end of the structure) measuring 0.36 m in height by 0.26 m in width. It is likely that a further sluice was present along the crest of the wall, but this had eroded away. The maximum surface of the basin is 12,320 m^2, with an estimated volume capacity of 30,000 m^3. The general state of preservation of the asset was good, especially given its lack of maintenance, presenting only some minor wall subsidence, crest erosion, and plant overgrowth.

The dam lies within the boundaries of the '*Los Vencedores de*' Cajabamba Alta community located downstream at an altitude of 3,687 m a.s.l. Work on the dam's restoration was undertaken in close coordination with the community during the 2022 dry season (April–October, especially between August–September 2022). The community fully supported the project, particularly given the earlier failure of the Ricococha Baja dam project, which had destroyed a partially functioning pre-Hispanic dam to build a modern cement micro-dam in 2008, this dam failed that same year.

In a departure from the use of cement and steel, our restoration project embraced the use of traditional materials and techniques combined with adaptable modern materials such as geotextile, geomembrane, and a mechanical sluice. Our approach to the dam restoration combined established wall conservation methods allied to local knowledge and modern technological advances. Rehabilitation of the dam went through several distinct phases (Figures 15.4–15.6):

1 Archaeological and geophysical survey of the dam including photogrammetry of structure.
2 Cleaning of extant structure of overgrowth.
3 Template trace survey of walls using plastic sheeting.
4 Photographic and drawn profile record of cleaned structure.
5 Codifying of all stone elements.
6 Removal of clay–gravel core and loose stones.
7 Placement of geomembrane (1.5 mm thickness) and geotextile (300 kg/m^2) in between inside and outside walls.
8 Recompositing of the dam core, using a compacted mixture of clay (2 parts), sand (1 part), earth (0.5 part), and gravel (0.5 part).
9 Conditioning of original sluice area to insert modern mechanical sluice mechanism.
10 Emplacing of loose stones and collapsed stones to achieve the original height of structure (Figure 15.5).
11 Final recompositing of crest core, stone crest and placement of top sluice at the level of dam crest (Figure 15.6).

The whole restoration process was recorded. Finally, it is important to note that in accordance with best-practice conservation protocols, reversibility was a key factor guiding the dam restoration. Therefore, it will be possible at a later time to remove all modern additions should a better solution be found in respect to their

Figure 15.5 Restoring fallen rocks manually onto the dam crest, Ricococha Alta. Past Water Futures Project.

Figure 15.6 Fully restored Ricococha Alta dam. Past Water Futures Project.

purported functions; this includes the geomembrane, geotextile, and mechanical sluice.

15.4.2 Engaging and Working with the 'Los Vencedores de' Cajabamba Alta Community

The Ricococha Alta dam restoration project with the *'Los Vencedores de' Cajabamba Alta* community had a long gestation period dating back to the early 21st century when part of this research group started working in the area (Herrera, 2005; Lane, 2006a). At the time, archaeological surveys and excavations centred on teasing out the mechanics of late pre-Hispanic (AD 1000–1532) polities in the region based around the social and economic underpinnings of agro-pastoralist communities (Lane, 2006b; 2007). Even then, the importance of water in conditioning the parameters of pre-Hispanic society was a salient issue. A key result of our research was the identification of numerous abandoned hydraulic management systems integrating soil and water conservation and usage across whole watersheds (Lane, 2009). In essence, *installed capacity* that could be reconditioned for use in the present day.

The easiest of these abandoned technologies to recover were the water dams given their increasing ubiquity as a quick, tried-and-tested modern response to water insecurity and scarcity in rural Peru (Autoridad Nacional del Agua, 2015). The need for additional water by rural communities in the area was made patently clear every time we visited the region. In this regard, the plight of the Cordillera Negra is particularly germane to hydraulic capacity building, given its drier environment in the past and its ever-increasing aridity in the present (Branch et al., 2023). Nevertheless, the shortcomings of many modern Central Andean micro-dams are equally plain to see. Of 13 dams in our study area, three of them were pre-Hispanic and were partially working, while the rest are modern (>30 years old), of which only one was continuing to function (Llosa Larrabure, 2008). Some like Ricococha Baja, the downslope twin to the recently restored Ricococha Alta dam, were catastrophic failures of planning, execution, and effectiveness. At Ricococha Baja, a new concrete micro-dam was built during 2008, effectively destroying its pre-Hispanic precursor. It lasted six months in operation, as cost-cutting and sub-standard materials, combined with indifferent engineering expertise, led to the dam leaking extensively during the first rainy season.

It was the failure of this modern dam, built with the *'Los Vencedores de'* Cajabamba Alta community's enthusiastic support and with what had been optimistically termed '*materiales noble*' (prestige material), such as cement and steel rods, that provided the local social conditions for the restoration of a pre-Hispanic dam. Given the modern dam fiasco, the *'Los Vencedores de'* Cajabamba Alta community was much more open to entertaining the possibility of rehabilitating an older dam, using lowly regarded *'materiales base'* (base materials), such as locally available rock, gravel, and clay, combined with a necessary upscaling of local construction knowledge, such as that used on traditional canals, to restore Ricococha Alta.

Restoration of the Ricococha Alta dam was coordinated closely with the village authorities and community members, taking a little over a month to complete. The work by the community was paid at a rate commensurate with a day's field labour. While restoration works were overseen by a team of trained restorers and archaeologists, the community provided the necessary construction know-how, including the location of suitable clay and gravel sources. The project foreman was also a community member with extensive experience in building stone and clay canals and walls. The placement of geomembrane, geotextile, and dam sluice was carried out under the guidance of the head restorer.

Throughout, local community buy-in on the project was crucial both for its execution and as a key component of ongoing maintenance of the dam. This last point is critical if the dam is to become an integral feature of the community. A series of guided workshops have helped instil in the community the need to adopt the structure within their yearly communal activities, for instance canal cleaning tasks which occur in April and October. Since the completion of the dam in October 2022, the dam has been regularly maintained and is working, having filled up three times during the 2022–2023 rainy season.

Still, one of the main challenges will be maintaining interest in the dam and its yearly maintenance by the community: times shift, community leaders change, and what seems relevant and important now might not be so in future. In the medium term, the research team will continue monitoring the dam and the extent of community engagement. We are nevertheless encouraged by the fact that since the successful restoration of Ricococha Alta other communities in the area have engaged with us towards restoring similar structures in their watersheds. To this end, we selected Weetacocha dam, belonging to the nearby community of Racratumanka as the next restoration candidate, with the work completed in 2024.

A further important challenge was finding the right grant with which to fund the Ricococha Alta project. Essentially, this had to wait for a recent significant shift in funding parameters, which now increasingly include climate change, social action, and heritage protection as essential considerations in their funding decisions. Alongside funding secured from the Gerda Henkel Foundation, the Harte Family Fund, and Rotary International, other bodies such as the World Monument Fund, the Mountain Institute – Peru, and McKnight Foundation have recently expressed interest in these types of rehabilitation projects demonstrating a shift from the preservation of sites towards site preservation coupled with measurable social impact.

15.4.3 Project Impacts

In the short term, the Ricococha Alta restoration project brought back into function a heritage dam with an estimated capacity of ~30,000 m^3. Since this restoration, the dam has been refilled three times, adding a further ~90,000 m^3 of water to the directly impacted 200 families from two adjacent communities (Cajabamba Alta and Putaca), while indirectly contributing water to a further 250 households downslope including Cajabamba Baja, among others. In total, this represents over 1200 people. This success has been instrumental in the community's adoption of the dam and its incorporation and continued maintenance during the twice-yearly chore of canal cleaning undertaken by the inhabitants of Cajabamba Alta.

The water produced by the Ricococha Alta dam is used primarily for downslope farming of subsistence crops such as maize and potatoes, but more importantly of cash crops such as peas (*Pisum sativum*) and further downstream avocado (*Persea americana*). Furthermore, the community of Cajabamba Alta will also use the extra water to grow alfalfa (*Medicago sativa*) as cattle fodder. The community maintains alfalfa-fed milk cows for their artisanal cheese factory; the cheese thus produced is sold in the nearby Caraz market. The milking of cows for cheese production is a key activity for women within the community, with the proceeds directly benefiting networks of female community members. There were at least two unintended proximate consequences of the dam restoration: firstly, an increase in bird numbers at and around the lake as they flock to this 'new' source of water; secondly, the emergence of new springheads downslope from the dam, as the newly filled dam basin replenishes underground geologic water aquifers.

In the medium to long term, the *Past Water Futures* project has provided a critical proof-of-concept pre-Hispanic dam restoration. Premised on the understanding that worsening climate conditions will negatively impact water sustainability and community resilience in the Andes, the successful restoration of an *installed capacity* asset such as the Ricococha Alta dam provides a community-led, low-carbon, low-cost, and low-maintenance alternative to modern cement micro-dams and water scarcity in the region.

The upscaling of this solution has enormous potential, with hundreds of ancient dams across the Central Andes available for rehabilitation. Such scaling relies on existing examples, such as the Ricococha Alta dam, that ably demonstrate to local communities the effectiveness of *installed capacity* restoration. In this regard, community buy-in for such projects is essential. Without the support from community members, the possibility of successfully restoring pre-Hispanic dams is compromised. Community participation is critical across the survey, building, and monitoring stages of the dam restoration. Only by making the dam theirs is success better guaranteed.

15.5 Reflections

The two case studies presented here amply demonstrate the forward thinking and long-term investment by rural pre-Hispanic communities in the face of past climate variability and socio-economic change. At present, given the widely recognised impact of the climate emergency in the Peruvian Andes, including debilitating economic and social effects of increased water shortage, land abandonment, drought, and crop failure, we believe that the deployment of cultural heritage data, and the utilisation of indigenous knowledge and expertise, could improve water security and ensure the sustainability of rural farming practices in the future.

Local and national government institutes, and NGOs, are aware of the growing problems and have, since the mid-1990s, undertaken development programmes aimed at counteracting the effects of water scarcity. Primarily, the solution has been to build concrete micro-dams along headwaters of valleys, i.e., across the front of existing wetlands. Essentially, this mirrors the pre-Hispanic adaptations we have recorded in these two case studies. Indeed, in many examples, the new dam has been placed on top of a pre-Hispanic structure. However, modern concrete micro-dams have significant weaknesses: they are expensive, rely on non-local expertise, and have a functional lifespan of only 50 years.

By contrast, ancient micro-dams were the product of millennia-long engineering projects that integrated these technologies with the immediate landscape and ensured wetland management from the top of watersheds downwards to agricultural fields. They were based on local knowledge of raw materials and would undoubtedly have been easily maintained. Furthermore, our Ricococha Alta restoration case study demonstrates that in these regions, where the poverty rate is high, the restoration of ancient micro-dams can rehabilitate the wetland ecosystem and benefit local farming communities. Indeed, having clearly demonstrated the benefits of micro-dam construction at Antaycocha for pre-Hispanic agro-pastoralists living in the Chillón Valley, it seems sensible that future restoration work should also be conducted in other parts of the Peruvian Andes, which will have considerable benefits for local communities across the wider geographical area.

The restoration project was rooted in a respect for communities and local knowledge, allied to flexible modern engineering. We contend that modern engineering can only provide partial solutions to improving water security in the Peruvian Andes and that the marriage between past and present knowledge and technology delivers a better, locally informed answer to future water stress due to climate change. Given that ancient micro-dams are common to the Andean highlands, our success on this rural development project we believe provides a cheap, easily applicable, community-based solution to water scarcity across large areas of the Andes.

Acknowledgements

We are grateful to the inhabitants of the villages located in Ancash and Lima where the research was undertaken. The project was funded by a NERC small research grant NER/B/S/2001/00256 (Branch, Meddens), AHRC SWW DTP studentship AH/L503939/1 (Handley), British Academy grant KF2\100158 (Branch, Meddens), Academy of Medical Sciences grant GCRFNG\100411 (Branch, Meddens), Royal Academy of Engineering grant FF\1920\1\74 (Branch, Meddens, Vivanco, Walsh), and a Gerda Henkel Foundation Humanitarian Aid and Social Welfare Projects grant Z08/HS/19 (Lane).

References

Apaéstegui, J., Cruz, F. W., Sifeddine, A., Vuille, M., Espinoza, J.C., Guyot, J.L., Khodri, M., Strikis, N., Santos, R.V., Cheng, H., Edwards, L., Carvalho, E. and Santini, W. (2014). Hydroclimate Variability of the Northwestern Amazon Basin near the Andean Foothills of Peru Related to the South American Monsoon System During the Last 1600 Years, *Climate of the Past,* 10(6), pp. 1967–81.

Autoridad Nacional del Agua (2015). *Inventario de Presas En El Perú: Primera Parte – 2015*. Lima: Ministerio de Agricultura y Riego/Autoridad Nacional del Agua.

Branch, N., Ferreira, F., Lane, K., Wade, A., Walsh, D., Handley, J., Herrera A., Rodda, H., Simmonds, M., Meddens, F. and Black, S. (2023). Adaptive Capacity of Farming Communities to Climate Change in the Peruvian Andes: Past, Present and Future (Preliminary Findings of the ACCESS Project), *Revista de Glaciares y Ecosistemas de Montaña*, 8, pp. 51–67.

Chepstow-Lusty, A. J., Frogley, M.R. and Baker, A.S. (2019). Comparison of Sporormiella Dung Fungal Spores and Oribatid Mites as Indicators of Large Herbivore Presence: Evidence from the Cuzco Region of Peru, *Journal of Archaeological Science*, 102, pp. 61–70.

Equihua, M., Aldama, M.E., Gershenson, C., López-Corona, O., Munguía, M., Pérez-Maqueo, O. and Ramírez-Carrillo, E. (2020), Ecosystem Antifragility: Beyond Integrity and Resilience. *PeerJ*, 8, pp. 8533.

Espinoza Soriano, W. (2013). Etnia Guaylla (Ahora Huaylas), *Investigaciones Sociales*, 17(30), pp. 179–90.

Farfán, C. (2011). 'Arquitectura Prehispánica de Cantamarca-Canta', in: K. Lane and M.L. Davila (eds.) *Arquitectura Prehispánica Tardía: Construcción y Poder En Los Andes Centrales*. Lima: UCSS, pp. 273–338.

Flores O.J.A. (1980). 'Causas Que Originaron La Actual Distribucion de Las Alpacas y Llamas', in: L. Milliones and H. Tomeada (eds.) *El Hombre y Su Ambiente En Los Andes Centrales*. Osaka: Senri Ethnological Studies, 10, pp. 63–92.

Gleick, P.H. (1993). Water and Conflict: Fresh Water Resources and International Security, *International Security*, 18 (1), pp. 79–112.

Grana, L., Camargo, N., Zarbulin, M.A., Larrain, A., Velenti, G., Tejerina, M.E., Basso, M. and Quesada, M. (2024). Los Reservorios de Agua En Las Redes de Riego de Casabindo (Dpto. Cochinoca, Jujuy). Variabilidad Tecnológica y Manejo de Agua Ancestral, *Arqueología*, 30(2), 13408. DOI: https://doi.org/10.34096/arqueologia.t30.n2.13408.

Grimm, E.C. (1991). *TILIA and TILIA.GRAPH*. Springfield: Illinois State Museum.

Haug, G. H., Hughen, K.A., Sigman, D.M., Peterson, L.C., and Röhl, U. (2001). Southward Migration of the Intertropical Convergence Zone Through the Holocene, *Science*, 293(5533), pp. 1304–8.

Herrera, A. (2005). *Territory and Identity in Pre-Columbian Andes of Northern Peru*. Unpublished Ph.D. Dissertation, Cambridge: University of Cambridge.

Kanner, L. C., Burns, S.J., Cheng, H., Edwards, R.L. and Vuille, M. (2013). High-Resolution Variability of the South American Summer Monsoon over the Last Seven Millennia: Insights from a Speleothem Record from the Central Peruvian Andes, *Quaternary Science Reviews*, 75 (September), pp. 1–10.

Lane, K. (2006a). *Engineering the Puna: The Hydraulics of Agro-Pastoral Communities in a North-Central Peruvian Valley*. PhD, Cambridge: University of Cambridge.

Lane, K. (2006b). Through the Looking Glass: Re-Assessing the Role of Agro-Pastoralism in the North-Central Andean Highlands, *World Archaeology*, 38(3), pp. 493–510.

Lane, K. (2007). 'The State They Were in: Community, Continuity and Change in the North-Central Andes, 1000AD-1608AD', in: S. Kohring and S. Wynne-Jones (eds.) *Socialising Complexity: Structure, Integration and Power*. Oxford: Oxbow, pp. 76–99.

Lane, K. (2009). Engineered Highlands: The Social Organisation of Water in the Ancient North-Central Andes (AD 1000-1480), *World Archaeology*, 41(1), pp. 169–90.

Lane, K. (2011). 'Hincapié En Los Andes Nor-Centrales: La Presencia Inca En La Cordillera Negra, Sierra de Ancash', in: K. Lane and M.L. Dávila (eds.) *Arquitectura Prehispánica*

Tardía: Construcción y Poder En Los Andes Centrales. Lima: Fondo Editorial UCSS/ CEPAC, pp.123–70.

Lane, K. (2021). Engineering Resilience to Water Stress in the Late Prehispanic North-Central Andean Highlands (~600–1200 BP), *Water,* 13, pp. 1–29.

Llosa Larrabure, Jaime. 2008. 'Elaboración e Implementación de Un Programa Nacional de Adaptación al Cambio Climático, Con Énfasis En Zonas Seleccionadas de La Sierra Centro y Sur Del País'. Lima: CONCYTEC/OAJ – Contrato de Subvención N°064.

Lüning, S., Gałka, M., Bamonte, F.P., Rodríguez, F.G. and Vahrenholt, F. (2019). The Medieval Climate Anomaly in South America, *Quaternary International,* 508, pp. 70–87.

Meddens, F. M., Venegas, D.M.A., Branch, N.P., Schmitz, L.F., Meddens, B.J. and Pomacanchari, C.V. (2024). The Cloud People of Naupallacta; Middle Horizon camelid management in the 2 hinterland of the Chicha Soras valley, *Ñawpa Pacha, Journal of the Institute of Andean Studies*, pp. 1–34. DOI: 10.1080/00776297.2024.2399428

Ministerio de Desarrollo Agrario y Riego (2019). Principales Conflictos Por El Agua Detectados En El Ámbito Cuenca Pampas. *Consejo de Recursos Hídricos de Cuenca Pampas* (blog). www.ana.gob.pe/2019/consejo-de-cuenca/pampas/DAA.

Paredes-Beltran, B., Sordo-Ward, A. and Garrote, L. (2021). Dataset of Georeferenced Dams in South America (DDSA), *Earth System Science Data,* 13(2), pp. 213–29.

Patel, E., Weinthal, E., Dabelko, G., Bruch, C., Daly, J. and Behnke, N. (2022). Water and Conflict: A Toolkit for Programming. Washington D.C.: USAID.

Percero-Oubiña, C., Fábrega-Álvarez, P., Troncoso, A., Salazar, D., Hayashida, F., Borie, C. and Pino, M. (2016). Sistemas Agrohidráulicos En El Loa Superior: El Caso de Topaín, *Boletín de La Sociedad Chilena de Arqueología,* 46, pp. 23–42.

Proyecto Glaciares+ (2018). *Construcción de Diques Para La Cosecha de Agua En Las Lagunas Periglaciares*. Lima: CARE Peru.

Rein, B., Lückge, A., Reinhardt, L., Sirocko, F., Wolf, A. and Dullo.W. (2005). El Niño Variability off Peru during the Last 20,000 Years: El Niño variability off Peru during the last 20,000 years, *Paleoceanography,* 20, pp. 1–17. DOI:10.1029/2004PA001099

Salminci, P.M., Tchilinguirian, P. and Lane, K. (2014). Bordos and Boundaries: Sustainable Agriculture in the High Altitude Deserts of Northwest Argentina, AD 850-1532, *Journal of Anthropology and Archaeology,* 2(1), pp. 189–218.

Stansell, N. D., Licciardi, J.M., Rodbell, D.T. and Mark, B.G. (2017). Tropical Ocean-Atmospheric Forcing of Late Glacial and Holocene Glacier Fluctuations in the Cordillera Blanca, *Peru, Geophysical Research Letters,* 44(9), pp. 4176–85.

Stark, J., Guillén, S. and Brady, C. (2012). *Follow the Water: Emerging Issues of Climate Change and Conflict in Peru. CMM Discussion Paper No. 5*. Washington D.C.: USAID.

Swain, A. (2015). 'Water Wars', in: J.D. Wright (ed.) *International Encyclopedia of the Social & Behavioral Sciences (Second Edition)*. Oxford: Elsevier, pp. 443–47.

Vining, B. and Williams, P.R. (2020). Crossing the Western Altiplano: The Ecological Context of Tiwanaku Migrations, *Journal of Archaeological Science,* 113, 105046.

Wieland, M. (2010). *Life-Span of Storage Dam*. International Water Power and Dam Construction February, pp. 32–35.

Part IV

Overview

16 Experiences, Lessons, Appreciation

Steven Mithen, Mubariz Ahmed Rabbani and Maria Rabbani

16.1 Threats to Cultural Heritage

The starting point must be the many and serious threats to archaeological and historic sites throughout the Global South as highlighted in our case studies. Civil wars and international conflicts have a devastating impact on cultural heritage. This can be to buildings and monuments themselves, as in the Swat Valley (Chapter 7), is always on the infrastructure that maintains cultural heritage, as in Kurdistan and Sudan (Chapters 6 and 10), and decimates tourism in the wider region that would have otherwise supported governments and communities to invest in their cultural heritage, as in Jordan and Egypt (Chapters 3 and 9). As we write in November 2024, the aggressors in the ongoing conflicts in Russia–Ukraine, Israel–Lebanon–Gaza, Sudan, Syria and Myanmar are destroying the cultural heritage of their opponents at an alarming rate. Whether that is by deliberate intent or as collateral damage, its long-term consequences for rehabilitating communities in war-torn areas are profound.

Other forms of conflict referred to in this volume include: the terrorist activity in Egypt during the early 1990s that reduced tourist numbers and left a long legacy of high security, conducive neither to visitors nor researchers at Amarna (Chapter 9): the damage to churches and convent buildings in Mexico that arose from their use as barracks during the Mexican Revolution (Chapter 12); and the drug traffickers in Ligüiqui, Ecuador, that continue to create a climate of fear and inhibit collaboration between researchers and members of the community, (Chapter 13).

Although feelings of despair are not unreasonable, solace can be found in the achievements of cultural restoration and how it supports economic development, as illustrated in the Swat Valley and Iraqi Kurdistan (Chapters 6 and 7); also, from how tourist numbers can bounce back, as in Kurdistan, the Swat Valley and Egypt (Chapters 6, 7 and 9). Moreover, there is the extraordinary commitment of people such as Sami Elamin, still working to protect the cultural heritage at Jebel Barkal during Sudan's ongoing civil war (Chapter 10).

A second category of threats arises from population growth and economic development: the expansion of farming, sprawling urbanisation, industrialisation with their requirements of quarrying, road building, water extraction and so forth. These can damage known archaeological sites, destroy those that have yet to be

DOI: 10.4324/9781003491071-20

discovered and reduce the attractiveness of areas for tourism. Such threats have been noted in most of our case studies, being particularly emphasised for those concerning Jordan, Iraqi Kurdistan, Egypt and Tanzania (Chapters 3, 6, 9 and 11). They play a role in Peru where steel and concrete dams are often built over and destroy the ancient pre-Hispanic stone-built dams (Chapter 15). In Pangani, Tanzania, economic developments destroyed and damaged built heritage without any consultation with the Department of Antiquities or care for what was being lost (Chapter 11).

The threats from economic development create moral dilemmas for archaeologists who work in regions of poverty: why should we prioritise the preservation of an archaeological site that may be of little academic or cultural consequence over a family seeking to secure their food supply by irrigating and cultivating a previously undisturbed area of desert? Some projects have no doubt that human needs take precedence: those workings with the Hualcayán community have a policy to avoid elevating the value of archaeological concerns above the day-to-day needs of local people (Chapter 14). That must be correct, but can be complemented by considering how cultural heritage can be used as an alternative means of economic development. By design, most of the case studies have sought to promote heritage tourism, anticipating this will lead to greater care for archaeological and historic while also providing income and employment opportunities. Other projects focussed on the provision of training for later employment in the conservation and restoration of archaeological sites, exemplified by Sela in Jordan (Chapter 5).

A third category of threats comes from climate and environmental change, having either direct or indirect impacts on cultural heritage. Global warming is taking its toll throughout the world, either by making regions virtually uninhabitable, and certainly unsuitable for tourism, by heightened temperatures, lowered water tables and higher sea levels, or by increased frequencies of droughts, floods and storm events. Such changes and events can hasten the decay of archaeological remains, such as the mud-brick walls found at Beidha and Amarna (Chapters 4 and 9). Mitigating actions can have negative impacts on cultural heritage, as with installing the concrete and steel dams in the Andes in a bid to secure water supply in the context of retreating glaciers (Chapter 15).

The COVID-19 pandemic can be considered as a global environmental event. It impacted all projects represented in this volume: for some, it thwarted initiatives by reducing global travel to such a level that tourism fell below critical thresholds, such as those necessary to sustain new business initiatives in Faynan, Jordan (Chapter 3); for others, it redirected national resources and attention away from heritage, as at Beidha, Jordan, leading to the deterioration of its archaeological remains (Chapter 4). In most projects, COVID-19 directly impeded the work of the archaeologists themselves by preventing travel and engagement with the communities they intended to work with.

As with economic development, climate change and environmental events can provide incentives for cultural heritage projects. In Ligüiqui, Ecuador, for instance, it was the disruption to farming and fishing arising from recurrent El Niño events, a 2016 earthquake and the COVID-19 pandemic that provided the motivation to develop heritage tourism to build the economic resilience of the local community (Chapter 13). Similarly, in the Philippines, the COVID-19 pandemic prompted the Department of Tourism to place greater emphasis on sustainable and inclusive tourism (Chapter 8). Moreover, the various disincentives for global travel and hence increased uncertainties about international tourism encouraged government policies and investments for domestic tourism in which cultural heritage plays a key role (Chapters 2, 4 and 9).

16.2 Defining, De-Limiting, Working with and the Changing Composition, of Communities and Stakeholders

As noted in the introduction, communities can be difficult entities to define and engage with. It is worth repeating Gould and Burtenshaw's remark that communities have individuals who *'may live together harmoniously, or not; they may have parallel economic interests, or not; and their identification with the heritage itself may range from profound to indifference'*.

That was certainly the case within several case studies in this volume. The projects in Faynan, Jordan, for instance, suffered because the Department of Antiquities had built the museum within the village of one tribe, where members of other tribes felt uncomfortable and may have been unwelcome, preventing the use of the museum as a community cultural hub (Chapter 2). The same tribal divisions contributed to unsurmountable tensions within the women's cultural association, despite initial wishes from the women, and the whole community, to have a non-tribal organisation (Chapter 3).

Community is especially hard to define and/or delimit when dealing with religious heritage in a region of diverse beliefs, especially when authorities adopt a secular approach. At Santa Cruz Mixtepec, Mexico, for instance, the restoration project prioritised working relationships with the municipal authority, whose cooperation was essential (Chapter 12). By doing so, however, it neglected other stakeholders, notably the parish priests, persons who live outside the municipality but who use the church and non-Catholics who have no representation within the municipal authority. This illustrates a much wider issue of how to manage the secular interests of heritage tourism and with the ongoing spiritual roles of religious buildings. The need to equip religious leaders with the knowledge and tools for how to steward religious buildings as they play a role in heritage tourism was noted for Cagayan, Philippines (Chapter 8).

One of the challenges is that communities are always changing, especially during and immediately following conflict. Iraqi Kurdistan, for instance, became a refuge for minority groups who fled from neighbouring territories, notably from Daesh between 2014 and 2017 (Chapter 6). Promoting a specifically Assyrian heritage within the resulting ethnically diverse population risks favouring the needs and values of some parts of the community over those of others. Similarly, in the Swat Valley, Pakistan, where contemporary communities might disapprove or even resist the restoration of cultural remains representing previous ideologies (Chapter 7). The impact of people displaced by conflict was especially demanding for the Jebel

Barkal project in Sudan because of the large numbers who relocated to its vicinity. Its local community dramatically expanded in size and diversity, causing the project to consider how it can work equitably during such crisis (Chapter 10). The population of the Hualcayán community doubled during the first fifteen years of the project (Chapter 14).

The challenge of defining the relevant community is matched by that of identifying the cultural heritage stakeholders and then finding the capacity to consult with them and the means to either resolve or select from their conflicting priorities (e.g. Chapters 8, 10 and 13). There is always, as Caroline Lising and Grace Barretto-Tessero remarked regarding the Philippines, 'an intricate web of stakeholders, characterised by diverse interests and expertise'. Stakeholders do not always consult or collaborate with each other, as with the UNESCO project that installed useful drainage at Beidha in a manner that damaged standing remains (Chapter 4).

Archaeologists can be confronted with local heritage management practices that conflict with their project requirements, as in Faynan with the Department of Antiquities policy of not allowing sales outlets within their museums (Chapter 3), and for the Amarna project regarding the building of walls around particularly threatened features (Chapter 9). As noted in Chapter 5, the policy of erecting fences around archaeological sites in Jordan created a sense of alienation of the local communities from their cultural heritage. To gain official status as a tour guide in Jordan requires a level of higher education, but this has often not been available or not achieved by those living next to archaeological sites who nevertheless have enthusiasm, detailed knowledge about and care for the remains (Chapter 4).

Stakeholders and supporters are always in flux. Changes in personnel can have negative impacts on projects, especially after a relationship had been carefully nurtured. The Uzikwasa NGO in Tanzania had been highly supportive of cultural heritage initiatives until a change of leadership altered its priorities, turning a heritage centre into a restaurant and compromising the sustainability of other initiatives (Chapter 11). In Jordan, a seemingly constant change of senior personnel in the Department of Antiquities between 2017 and 2022 meant proposals had to be repeatedly reintroduced for their approval. The sudden and unexpected departure of the highly supportive Minister for Culture and Tourism in 2019 (to become an overseas ambassador) removed one of the most influential champions of Faynan projects – fortunately she has returned to that ministerial role in September 2024 (Chapter 2). Despite a positive record of maintenance for the restored Ricococha Alta dam, Kevin Lane and his colleagues fear for its future, noting how 'times shift, community leaders change, and what seems relevant and important now, might not be in future' (Chapter 15).

Any project is likely to have a different relationship with different stakeholders and elements of the communities they work with. The ideal is to work in partnerships, enabling the co-production and ownership of knowledge and outputs. The CONCH project working with communities at Pangani and Kilwa in Tanzania (Chapter 11) had a strong commitment to such co-production but recognised it as not always possible, noting that 'hierarchies between professionals and local communities are often deep rooted'. Similarly, while the ideal is for initiatives to develop bottom-up, those of a significant scale, which require expert knowledge and significant resources, inevitably derive from top-down. Relationships with stakeholders and communities will always be in constant flux because the projects themselves are always having to change. While they may start with well-defined aims and plans for actions, they invariably need to respond to events, unexpected problems and new opportunities when in the field.

Archaeological work itself can impose difficulties for communities, creating newfound tensions and disputes between families, and between the community and the archaeologists. In the Hualcayán community of Peru, for instance, those families whose agricultural land was revealed as containing significant archaeological sites came under pressure from other families to avoid destroying the archaeology, while also being reluctant to give up their own land for use as an alternative area for ploughing (Chapter 14). Similarly in the Swat Valley, Pakistan, where families may be reluctant to allow heritage projects on their own land (Chapter 7). That is because if archaeological remains are found and designated as a site of cultural significance, the land might be acquired by the state for further protection of the archaeological remains. Because buffer zones are required around notified archaeological sites, construction projects are sometimes launched in their vicinity prior to the process of notification for fear that the land might be later expropriated by the state.

16.3 Material Infrastructure, Interventions and Museums

None of the case studies attempted or favoured the construction of imposing infrastructure as a means to protect archaeological sites. As noted for Jordan (Chapter 4) and Ecuador (Chapter 13), these can become white elephants, acting to hide the cultural heritage; they have high maintenance costs and experience indicates they may neither benefit local communities nor attract tourists. The costs to erect them are better spent on a portfolio of smaller, more discrete initiatives, which will also avoid the environmental impacts and carbon emissions arising from such constructions.

The establishment of museums is a perquisite for most initiatives concerning heritage tourism – visitors expect them, and museums indicate there is valuable heritage to explore. As emphasised in several case studies, museums can deliver a dual function of (1) providing information and experiences to tourists and (2) engaging local communities with their heritage, especially if they become their hubs for cultural activities. That is being achieved at Hualcayán, Peru (Chapter 15), and by the Amarna Project, Egypt, which has sought to repurpose a visitor centre designed for outside visitors into a heritage and community hub to serve local needs (Chapter 9). Achieving the dual functions remains an aspiration for Faynan, Jordan (Chapter 2), and Ligüiqui, Ecuador (Chapter 13).

Walking trails are a means of introducing visitors to archaeological sites in their landscape and blending interests in heritage and hiking; they can also be of value for informing the local community about their cultural heritage, as planned in Ligüiqui, Ecuador (Chapter 13). The smaller-scale interventions they involve can, however, be problematic. The erection of signs and information boards is a relatively low-cost way to draw attention to archaeological sites, whether on walking trails as in Faynan (Chapter 2), or to demarcate the location of sites to avoid unintended disruption (Chapter 9). The dilemma is that information boards and signs can also be obtrusive and detrimental to the natural environment, which also needs to be preserved for effective tourism. Moreover, information panels and signs frequently became damaged through either deliberate vandalism or accidental events, while images and text can fade due to the intensity of light, heat and cold – despite the guarantees of suppliers (Chapters 2, 4, 9 and 11). The consequence is that they either become an eyesore in the landscape or require regular maintenance that imposes recurrent costs, extending beyond the duration of project funding. An alternative is to minimise such signs and embrace the use of digital technology by using smartphones and QR codes, as has become standard at heritage sites in prosperous regions in the Global North. Rebecca Bria and her colleagues are planning something even more ambitious at Hualcayán involving smartphones to trigger 3D experiences from codes on archaeology-themed sculptures (Chapter 14). The problem in many regions of the Global South is that temperatures, dust, vandalism, accidental damage and problematic networks make it unlikely that such digital technology would survive for long. As was employed at Kilwa, Tanzania, low-tech solutions will often have the most positive outcomes (Chapter 11).

The restoration and conservation of archaeological sites involves another type of material intervention. As demonstrated by the work of the Italian Mission in Swat, the requirement is to ensure that local materials are used, ideally the same that had been lost (Chapter 7). This maintains the authenticity of the monuments, reduces dependency on imported materials (with their attended costs and environmental impact) and will enable traditional skills to be maintained within the communities. A purist approach towards traditional materials may, however, be unwise. The success of the Ricococha Alta dam reconstruction depended on the combined use of traditional materials and techniques with adaptable modern materials such as geotextile, geomembrane and a mechanical sluice (Chapter 15).

16.4 Training and Education

The provision of education and training is integral to all case studies. Where this has not been delivered, there are compelling arguments for its need, notably for cultural resource management (CRM)in the Philippines and Sudan (Chapters 8 and 10) and outreach training by Visitor Centre staff at Amarna Egypt (Chapter 9). The training delivered by the projects falls into two main categories. Two projects, Sela in Jordan (Chapter 5) and Archaeology, Community, Tourism – Field School (ACT) in the Swat Valley, Pakistan (Chapter 7), focused on training the local workforce in methods of not only archaeological excavation but also restoration and conservation, both with impressive programmes. Sela worked with the Jordanian Department of Antiquities to devise profiles for professional and technical CRM staff, established training programmes and could demonstrate their success by the numbers of trainees who secured employment. The ACT project noted the role of certification for trainees, how increasing the number of skilled workers could reduce dependency on external professionals and how training provides transferable skills to a workforce – useful once project funding has ceased. The ACT project has a particularly impressive vision of skills cascading down through a workforce, as those trained become the trainers for the next generation, enabling the community to become self-sufficient.

Other projects delivered training focussed on archaeological methods and skills, those of surveying, excavation, geophysics and so forth. In Tanzania, these were provided to students and teachers, intending the latter to bring them into their schools and institutions (Chapter 11); in Ecuador, the focus was on the youths who joined their Young Guardians programme (Chapter 13), while in Iraqi Kurdistan, training was provided to the existing employees of the Directorates of Antiquities. Other types of training included that of tour guides (Chapters 4, 10 and 11) and in handicraft skills and business practice in Faynan (Chapter 3).

One of the challenges lies not with delivering the training itself but with changing self-perceptions so that those receiving the training can operate independently once the training has ended. In Faynan, Jordan, the women who wished to make and sell handicrafts to tourists received payments while training in craft skills and business (Chapter 3). They struggled to accept that such payments could not continue once the training was finished (when project funding had ceased), requiring them to generate income via selling their products: a long history of aid projects in the region had instilled a culture of dependency that could not be replaced in the duration of the project. A similar change of self-perception was required in the Swat Valley. Local people were trained as heritage tour guides, who then established their own organisation (Chapter 7). They had difficulty, however, moving from the status of wage labourers to that of self-employed workers and entrepreneurs, that required a change in their self-identity that had been inherited for generations.

Engaging with schools is integral to numerous of the case studies. The motivation is to raise awareness within children of their cultural heritage via broadly defined educational activities; anticipating this will instil values about respecting archaeological remains and inspire some towards future training and ultimately employment in the heritage sector. As Maria Elena Ronza writes in Chapter 5, introducing children to history in a playful way can 'ignite children's curiosity and imagination, laying the foundation for a lifelong journey of discovery and appreciation for cultural heritage'.

In Faynan, Jordan, the information boards suffering the most vandalism were those close to the walking routes to schools. The extent of damage was reduced following a programme of school visits to the museum, craft activities involving making mosaics and refitting pottery sherds and production of books about Faynan's archaeology for use in classroom and at home (Chapter 2). Sela has become the champion of such work in Jordan, delivering such activities in schools and at seasonal camps, producing heritage themes in children's books and turning them into animated films (Chapter 5). The Amarna Project did likewise, while also providing site tours and working in parallel with UK schoolchildren to enhance their understanding of both ancient and contemporary life in Egypt (Chapter 9). Site visits and activities were organised for children by the projects in Iraqi Kurdistan, Sudan and Peru (Chapters 6, 10 and 14). The CONCH project in Tanzania not only arranged walking tours and site visits for schoolchildren and their teachers, but also developed an extensive set of teaching materials for use in schools. These included comprehension exercises covering history and architectural heritage, wall posters, a map quiz and board game, with training events and workshops for the teachers (Chapter 11).

16.5 Funding Scales and Project Durations

The case studies varied dramatically in their level of funding (often not made explicit) and duration. Those in Iraqi Kurdistan and the Swat Valley were supported by multiple awards from private foundations, government bodies and/or academic institutions and have a relatively long duration (Chapters 6 and 7). Others, such as those UK-led projects in Faynan, Amarna and Tanzania (Chapters 2, 3, 9 and 11), relied on relatively small and short-term awards for 'impact work' from various UK funding schemes, extending over a three-year period at most, although the projects built on long-term research commitments in those locations; other projects fall somewhere in between. Whether large or small, all projects had relied on funds from multiple sources. Unfortunately, the effort in securing such awards (and the failures along the way) is often neglected when such projects undergo evaluations.

Inevitably, the scale of funding influences the aims, methods and what can be achieved. While few will contest that 'bigger is better', the dangers of investing in large material infrastructure that cannot be maintained over the long term have already been noted. Moreover, managing and spending large budgets in rural areas of the Global South can often be challenging when systems of financial accountability are less than one would desire.

The difficulty with short-term awards is that the expectations of the funding bodies are unrealistic. In the UK, for instance, academics are encouraged to draw on their research to undertake impact work, which aims to enhance economic growth, development and social and individual well-being. Between 2014 and 2021, funding for impact was provided by the Newton Fund and the Global Challenges Research Fund (GCRF) schemes, often supplemented by university-based impact award schemes. Most Newton and GCRF awards were for science-based projects; those awarded for the humanities were relatively small and had a restricted time frame for completion.

As illustrated by the challenges described in our case studies, while much activity can be undertaken within the typical three years of such a project, and positive outcomes achieved, whether such outcomes will continue over the longer term is unknown. As Rebecca Bria and Eric Vásquez wrote about being absent from the Hualcayán community during the COVID-19 pandemic years of 2020–2022, 'Did anyone in Hualcayán even think about archaeology once we left?' (Chapter 14). Such sentiments are likely being shared by other project leaders during periods of absence from their field sites. At most, Newton and GCRF-funded projects (and equivalents from other countries) seem able to make a short-term difference when operatives are on the ground working with communities but are unable to have confidence that any of their contributions will be sustained over a longer time frame. This is indeed the conclusion reached by the final evaluation of the Newton Funding Scheme (BEIS 2022),[1] which could only describe 'sustainability' in terms of how some projects have secured follow-on funding and continue to produce publications because none of the projects surveyed could provide evidence that they had made a significant and sustained difference to the communities they worked with.

Long-term projects also have their risks. Another insight from Rebecca Bria and Eric Vásquez is that the longer you work in a community the higher the stakes become. A two-year project with clear goals and outcomes for all parties can end comfortably, but over the longer terms expectations on both sides are elevated. That occurred in Faynan because it received a succession of internationally funded archaeological research projects from the 1980s up until 2010, with each of these employing workmen, providing valuable income to the community. An expectation arose that such funding was endless, and consequently, any visit that I or other archaeologists now make is assumed to be in preparation for further excavation and employment.

Nevertheless, long-term engagement with communities must, to some extent, be a perquisite for a successful collaboration. Building relationships of trust cannot be rushed. Community engagement in the Jebel Barkal project began, as it must anywhere, with 'meeting people for tea, introducing ourselves and the project, listening about people's daily lives and making observations, attending cultural events and building relationships and trust to establish a strong basis for future seasons (Chapter 10). Rather than imposing restrictions on how the archaeological sites can be treated, the project worked with the community to find mutually agreeable solutions, such as negotiating with municipal waste disposal for trash disposal that would have otherwise ended up on the site. The value of 'going slow' for building and maintaining relationships of not only trust but also care is sacrosanct for those working within the Hualcayán community in Peru (Chapter 14). That took them beyond the boundaries of archaeology to assist with a variety of community problems ranging from completing new school rooms to maintaining dirt roads after heavy rain and supporting individuals in their personal challenges.

16.6 Measuring Success

In our experience, few academics can allow themselves the time to match the commitment of Rebecca Bria and her colleagues to the Hualcayán community. They are usually undertaking their cultural heritage projects as part of a research portfolio with other projects to deliver, along with teaching and administrative commitments at their institutions (not to mention finding a work–life balance for their own well-being). Moreover, they are required to report to their funding bodies and employing institutions who wish to know how funds have been spent and what achievements have been made; those bodies may not appreciate the value of 'hanging out with everyone from tea ladies to footballers', as Rebecca Bradshaw and colleague described their work in Sudan. Funding bodies and employing institutions want quantitative data: How many new visitors to a museum has a project generated? How many new jobs have been created? To what extent has a project enhanced gender equality and reduced poverty? Such measurements are usually impossible for projects to derive, whether they were of short- or long-term duration. As with the Amarna and Faynan projects, funding has normally run out before it remains feasible to begin collecting any such data (Chapters 2 and 9). Moreover, when estimates for such changes can be derived, attributing them to the work of a single project will always be problematic because other factors are in play. The ACT project in the Swat Valley, for instance, was able to report on a significant increase in tourist numbers as the project progressed, but recognised political developments in the region were also critical, notably the start of the China–Pakistan Economic Corridor (Chapter 7).

Success is most easily measured when there is a discrete and measurable goal. The most explicit in these case studies has been the restoration of the Ricococha Alta dam in Peru and its hugely beneficial impact on local communities (Chapter 15), similarly for failure, in these case studies that of establishing a women's handicraft business in Faynan (Chapter 3). Even with these projects, however, declaring their success and failure is problematic. We have already noted how the success of the dam will depend on constant maintenance by the community that cannot be guaranteed, while at least some of the women who received training in crafts and business in Faynan may use their new skills in other contexts for their own benefit. Indeed, one of the most able women left Faynan soon after being trained in accounting, armed with her new skills to find employment elsewhere. Moreover, positive project outcomes may only arise long after final reports have been submitted, such as when those children whose curiosity and imagination were ignited by joining a dig, washing pots or hearing a story about the past, reach adulthood and become cultural heritage professionals.

16.7 Final Thoughts

Despite seeking to make this volume representative of the field in terms of scope, scale and the diversity of our selected case studies, we do not claim it to be an exhaustive or conclusive piece of work. There are other ongoing projects at archaeological sites that are embracing community engagement and supporting sustainable tourism-related practices, reflecting a growing commitment by scholars to integrate these practices into their research projects. In Iraqi–Kurdistan, for instance, there is the Sirwan Regional Project (University of Glasgow, UK) and the Central Zagros Archaeological Project (University of Reading, UK). The University of Durham, UK, has active cultural heritage projects in southern Nepal, as does the University of Liberal Arts Bangladesh at the site of Bhitargarh.

This volume highlights the profound interconnectedness between cultural heritage, community engagement and sustainable tourism. Its diverse case studies reveal both the challenges and opportunities in leveraging heritage for broader societal benefits. Despite the complexities we have highlighted, they demonstrate the transformative potential of heritage work – not only in preserving the past but also in shaping resilient futures for local communities. As we navigate a time of unprecedented global change and uncertainty, these reflections invite continued dialogue and meaningful action between all parties involved; they require all stakeholders and especially those in authority to embrace adaptive, inclusive and long-term strategies that place local communities at the heart of heritage management and sustainable tourism.

Note

1 The UK's Department for Science, Innovation and Technology and Department for Business, Energy & Industrial Strategy published their evaluation of the Newton Fund with supporting evidence at www.gov.uk/government/publications/newton-fund-final-evaluation-and-supporting-evidence

Index

Note: Page numbers in *italic* refers to Figures.

FSC
www.fsc.org
MIX
Papier | Fördert gute Waldnutzung
FSC® C083411

Kolibri 360 GmbH
Ferdinand-Jühlke-Straße 7
99095 Erfurt, Deutschland
produktsicherheit@kolibri360.de